# Maker Pro

Altman, Baichtal, bunnie, DiResta,
Dyba, Gauntlett, Gentry, Hord,
Jankowski, Klingberg, Kravitz,
Krumpus, Meno, Petrone, Smith,
Solarz, Tremayne, Wang, and Wolf
**Foreword by Joey Hudy**

MAKER MEDIA™

SEBASTOPOL, CA

# MAKER PRO

by Altman, Baichtal, bunnie, DiResta, Dyba, Gauntlett, Gentry, Hord, Jankowski, Klingberg, Kravitz, Krumpus, Meno, Petrone, Smith, Solarz, Tremayne, Wang, and Wolf **Foreword by Joey Hudy**

Published by Maker Media, Inc., 1005 Gravenstein Highway North, Sebastopol, CA 95472.

Maker Media books may be purchased for educational, business, or sales promotional use. Online editions are also available for most titles (*http://safaribooksonline.com*). For more information, contact our corporate/institutional sales department: 800-998-9938 or *corporate@oreilly.com*.

| | |
|---|---|
| **Editor:** Brian Jepson | **Indexer:** WordCo Indexing Services |
| **Production Editor:** Melanie Yarbrough | **Cover Designer:** Riley Wilkinson |
| **Copyeditor:** Phil Dangler | **Interior Designer:** Monica Kamsvaag |
| **Proofreader:** Sonia Saruba | **Illustrator:** Rebecca Demarest |

December 2014:     First Edition

**Revision History for the First Edition:**

2014-12-08:    First release

See *http://oreilly.com/catalog/errata.csp?isbn=9781457186189* for release details.

ISBN: 978-1-457-18618-9
[LSI]

# Contents

# Foreword

Hello, my name is Joe Hudy and I'm a Maker.

I started Making about four years ago. Back then I was just making things out of cardboard and old toys. I made things like trebuchets, coin pushers, door openers, and other cardboard-centered things. One day, my mother decided to call up a company named Elenco Electronics to see if we could get any Snap Circuits, which are basically snap-together circuits. On the receiving end of the call was a man named Jeff Coda. Jeff helped me become a better Maker. He sent me my first soldering iron and a couple of products to test and solder together. This was extremely helpful and enabled me to learn about the world of electronics.

After a while of soldering and learning more about electronics, Jeff asked my mother if we had ever been to a Maker Faire. We hadn't at this point, and my mom didn't even know what they were! Jeff urged her to take me, and eventually Jeff sent us tickets and said, "Now you have to go." At this time I was making my extreme marshmallow cannon, an air cannon that shoots marshmallows, and decided to exhibit it at the Faire.

When I arrived at my first Faire, I realized that this is where I belonged: in the Maker community. I could finally speak "normally," talking about Tesla and electronics. I had been introduced to a wonderful new world full of opportunity. Before coming to the Faire I had heard about Arduinos, and when I was there I realized that a large majority of the Makers there used them. Because of this, I persuaded my mom to buy me a Arduino, and that was the second best thing that ever happened to me—the first was going to the Maker Faire.

After my first Faire, an avalanche of events emerged. I started fundraising to go to the next Faire, and soon I was going to all of them—making great friends with the *Make:* magazine crew and other Makers. At this point I had created my own product using the Arduino—a 3 × 3 × 3 LED Cube Shield, which I had gotten into the Maker Shed, the official store of Make:.

After a while of touring Maker Faires, I got a call asking if I wanted to go to the White House Science Fair to represent *Make:* magazine, and of course I said yes. At the White House, I was exhibiting my Extreme Marshmallow Cannon (you can see it in Figure F-1) and it just so happened that I was in the right room at the right time and President Barack Obama entered the room, looking at the exhibitors. When he got to my booth, he asked if we could fire the cannon, and of course I said yes. In shooting the cannon, I became a Maker icon in a way. As a result of shooting a cannon with the President, a ton of opportunities emerged: going to all the large Maker Faires, giving speeches, traveling to Italy and China for Faires, landing an internship at Intel, and even getting invited to sit with the First Lady at the 2014 State of the Union, where I also got to meet President Obama for the second time.

All of this was possible because of the Maker community, and it is possible for you, too.

**Figure F-1.** *Joe Hudy. Photo credit: Julie Hudy.*

Joe Hudy (*http://lookwhatjoeysmaking.blogspot.com/*) is the youngest-ever employee at Intel. He's already a veteran Maker who has completed several successful projects and kits, including his 3 × 3 × 3 LED Arduino Shield, SMD LED Arduino Shield, and Marshmallow Cannon. His latest project is a 3D Body Scanner (*http://bit.ly/1oMjV59*), which he demonstrated at Maker Faire New York. Joey's slogan is: "Don't be bored...make something!"

# Introduction

There is the concept of the Dark Age among Lego fans—defined as the time when you lose your love of the little plastic bricks and turn your attention to seemingly more important exploits: getting a job, finding a partner, or gaining an education.

As your attention shifts, your Lego collection languishes in the basement—sometimes for years—before finding its way to Mom's garage sale or a cousin's house. People entering their Dark Age decide they don't have time for the frivolity of toys. Responsibilities trump hobbies, and Lego bricks are associated with immature pursuits.

All that time, your love of Lego never quite went away...and before you know it, it's back in full force. Maybe a friend or relative gives you a Lego kit for a gag gift. Maybe you spot a bucket of bricks at a yard sale, and can't resist. Somehow, the thought occurs to you that building a model might be fun again.

Thus, the Dark Age ends.

These adult builders discover that not only can they recapture the creative fun they enjoyed before, but that it's far better than they could have imagined—buoyed by an adult's sophistication and pocketbook, they can build projects that would have made their jaw drop as a kid. They build Rubik's Cube solving robots, re-creations of legendary battleships, sophisticated vehicles, and countless other challenging projects.

Making works much the same way. When we're kids, we're always making, scribbling, messing up, getting discouraged, experimenting, and learning. Our minds are sponges of technique and tool lore. I've employed tools that I saw my dad use once, and think clearly back to the methods he used. Kids participate in science fairs and build kits.

It's not too hard to see a connection between Making and playing with a creative toy. Building sets and modular electronics mimic Lego's snap-together elements and robotics add-on sets. But, as with Lego, there comes a time when we stop making and settle down to the business of being a grown-up. The chemistry set gets smashed under another box in the basement, while the soldering iron ends up unused at the bottom of a tool bin.

The Maker Dark Age ends the same way the Lego one does—flipping through a *Make:* magazine at the bookstore, or touring the local hackerspace. Giddiness ensues, as you take the first tentative forays into Making. You might purchase an Arduino starter kit or Raspberry Pi. You learn how to solder, how to program. Every project is a little more challenging than the last.

If you keep going, something odd happens. You construct a tool (say, an electronic project) and it occurs to you that it would be easy to make a whole bunch of them. It's not greed, nor do you expect to make a living at selling kits—at least at first. And yet, for some of us, the opportunity presents itself. Maybe you spend all of your free time doing fulfillment while holding down a day job.

Then you decide to quit and become a Maker Pro.

This book is a celebration of those who have gone beyond "the hobby" to make a living at it. Most of the authors and interviewees are self-employed or flirting with it. They sell products, but their spare time is spent learning about new hardware and software, and building beautiful projects to highlight what they've discovered.

Is this the future, where a legion of artisanal Makers can be relied upon to build our couches and lamps and clocks? I don't know, but I'm excited to be a part of this resurgence of handicrafts.

So: end your Dark Age, make something, grow, learn, and maybe someday, when all the stars are aligned, it'll be time to go pro.

— John Baichtal, November 2014

# Preface

## How to Contact Us

Please address comments and questions concerning this book to the publisher:

> Make:
> 1005 Gravenstein Highway North
> Sebastopol, CA 95472
> 800-998-9938 (in the United States or Canada)
> 707-829-0515 (international or local)
> 707-829-0104 (fax)

Make: unites, inspires, informs, and entertains a growing community of resourceful people who undertake amazing projects in their backyards, basements, and garages. Make: celebrates your right to tweak, hack, and bend any technology to your will. The Make: audience continues to be a growing culture and community that believes in bettering ourselves, our environment, our educational system—our entire world. This is much more than an audience, it's a worldwide movement that Make: is leading—we call it the Maker Movement.

For more information about Make:, visit us online:

> Make: magazine: *http://makezine.com/magazine/*
> Maker Faire: *http://makerfaire.com*
> Makezine.com: *http://makezine.com*
> Maker Shed: *http://makershed.com/*

We have a web page for this book, where we list errata, examples, and any additional information. You can access this page at *http://bit.ly/maker-pro*.

## Safari® Books Online

 *Safari Books Online* is an on-demand digital library that delivers expert content in both book and video form from the world's leading authors in technology and business.

Technology professionals, software developers, web designers, and business and creative professionals use Safari Books Online as their primary resource for research, problem solving, learning, and certification training.

Safari Books Online offers a range of plans and pricing for enterprise, government, education, and individuals.

Members have access to thousands of books, training videos, and prepublication manuscripts in one fully searchable database from publishers like O'Reilly Media, Prentice Hall Professional, Addison-Wesley Professional, Microsoft Press, Sams, Que, Peachpit Press, Focal Press, Cisco Press, John Wiley & Sons, Syngress, Morgan Kaufmann, IBM Redbooks, Packt, Adobe Press, FT Press, Apress, Manning, New Riders, McGraw-Hill, Jones & Bartlett, Course Technology, and hundreds more. For more information about Safari Books Online, please visit us online.

## Acknowledgments

Thanks to Alex Dyba for the idea, and to Brian Jepson for making this book possible. There wouldn't be a book without the great contributors: Mitch Altman, Jimmy DiResta, Alex Dyba, David Gauntlett, Eri Gentry, Mike Hord, Joe Hudy, bunnie, Tito Jankowski, Rob Klingberg, Sophi Kravitz, Michael Krumpus, Joe Meno, Emile Petrone, Zach Smith, Susan Solarz, Wendy Tremayne, Akiba, and Adam Wolf.

A special thanks to my friend Riley Harrison of Twin Cities Maker for the trebuchet demo!

# The Art of Unemployment

WRITTEN BY **WENDY TREMAYNE**

Marisa and I had just met. After a quick lunch in a new mom-n-pop Asian fusion restaurant, we went to my place and chatted on my thrift-shop-scavenged, mid-century modern couch. I picked it up at the local consignment shop and will resell it back to them one day. We spent the rest of the afternoon swapping stories of the ways that we learned to free ourselves from full-time employment.

Marisa came to New Mexico so that we could have this conversation and provide background for a film she is making about folks who live alternative lifestyles and reside in trailers and mobile homes. My partner Mikey and I share more in common with Marisa than living in a renovated, ancient mobile home. In spite of our living on different ends of the country, common to our lives is that our lives are our jobs, and it was not always this way. I poured her a second glass of homemade ginger ale while agreeing that for me, having a job would be expensive.

I told Marisa how the lifestyle that Mikey and I currently enjoy got its start. In 2006, we were living in New York City, where we defined our lives as a patterned loop of working really hard to make money to buy things that we probably could have made better and more responsibly if we had made them ourselves. Our jobs, when looked at this way, seemed expensive to us because they took up all the time that we could have used to make what we bought. We felt dragged down by the cost to life of our consumer lifestyle, and yet we could see no clear way out of it without making big changes. It was at this time that I ran into a quote by an Indian philosopher. It stung when I first heard it, and I knew then that we were going to change the way we were living. "It is no sign of wellness to be well adjusted to a sick society." (Krishnamurti)

After quitting our corporate jobs—I was a creative director at a marketing firm and Mikey worked in IT for a Wall Street bank—we pledged to make our own goods from what could be found in the waste stream or from nature, rather than demand that new raw materials be harvested for our consumption. We sought abundance rather than wealth. Then we moved to a small town located in southern New Mexico called Truth or Consequences, with no idea what we would do for money. It was sufficient that we agreed to figure it out and agreed to live by the ideals that had inspired us to turn our lives around. One of those ideals included the belief that we are essentially creative. We suspected that the specialization required by our careers masked this fundamental fact.

**❝❞**

We treat money as the low road, especially in exchanges with friends who make things. Once you make something, you have it to barter, trade, and gift, and this is worth more than money. Barter comes second to gifting and is used in exchanges with acquaintances. Money is saved for total strangers and represents the absence of any relationship.

Realizing that I'd been rambling on and on about myself, I interrupted my own flow of thought to ask, "So Marisa, why is your trailer parked in Massachusetts?" She told me that, like my manufactured home, her trailer was made in 1967 and that it is in Massachusetts simply because that is where she and her sweetheart are from. They are fond of the uncommodified and not-yet gentrified little town in which they live. This statement set off a enthusiastic round of agreement between us about the way that a low cost of living is essential to living a decommodified life. This usually requires living in places far away from big cities. Truth or Consequences (T or C), New Mexico, could easily be called impoverished if one were to judge it by quick glance. But those of us who live here enjoy the benefits of freedom from a full-time job and time to grow high-quality organic food (and a year-round growing season), ferment exotic foods (things like cheese, wine, yogurt, tempeh, and kimchi), make DIY biodiesel, forage building materials, and enjoy an unblemished natural environment. You can see some of our projects in Figures 1-1 through 1-3. What we enjoy the most, though, is time to discover and work on our own creative projects. I told her with excitement that eight years into our adventure we actually do make things better and more responsibly than industry could have, just as we envisioned when we decided to leave the city.

I shared renovation techniques; our home was completed at a cost of ten dollars per square foot, using waste materials. "We saw the mobile home as a way to learn to use power tools, figuring that we couldn't make it worth less even if we failed." "A winning bet!" Marisa cheered. I told her that this was just one of the ways in which I learned to be a maker of things. After the mobile home renovation, Mikey and I made biodiesel, built a paper-crete dome (a mix of repulped paper and cement) to house a photovoltaic (PV) solar array that we installed; learned to make plant medicines; grew, fermented, and processed wholesome food; built homebrew electronics; made textiles; and provided ourselves with an astonishing variety of high-quality domestic goods like the wine shown in Figure 1-1.

**Figure 1-1.** *Wendy Tremayne and her partner Mikey Sklar bottle wine made from a kit. Photo credit: Mikey Sklar.*

Marisa's freedom, like my own, is linked to an ever-shifting strategy that provides the little income a homesteader needs. We agreed that reducing one's cost of living is the first step to freedom. I shared that it was also important to me that I begin this lifestyle change with a gift: after all, I was about to test a theory that the world is essentially abundant and was soon to rely on that abundance. Why not start by giving? I put the first successful project that I created once I was free of a job (the website (*http://swaporamarama.org*)), into the commons so that it could be shared. I went on to tell her about the eBay store that we use to sell treasures found at yard sales, and revealed that we have a small cottage industry through which we sell botanical medicines I make from plants found in the Chihuahuan Desert, and homebrew electronic gadgets that Mikey makes. "We make extra of the things we need ourselves and sell them online." Mikey's gadgets, for example, include a temperature controller that we used to convert a chest freezer into a refrigerator, thereby reducing the power draw of the fridge by 10× giving us power to run an air conditioner and welder on what our PV solar array produces. We also use the temperature controller to ferment foods and beverages. His homebrew battery desulfator recharges nearly all varieties of dead batteries. We no longer purchase new batteries, nor do we send dead ones to the landfill. "We sell gadgets and other products for a postconsumer life in our online store," I told her, as she scribbled down the URL.

Marisa told me about e-courses she creates that have, at times, generated thousands of dollars at once during the week she announces them. "It's passive income after that," she says with a sense of accomplishment and astonishment. I jot the words "start an e-course" on a piece of paper. She too runs an online store. Her Etsy store is based on a formula of listing 500 items of vintage clothing and then letting the store run until the inventory is too low to generate much income. Then she lists another 500 items. I admitted to her that nearly everything that decorates my home is listed on eBay. "I'm not attached to things," I said. "I like when they come and go. It makes my life more interesting and everything came from the waste stream so there's no worry about the destruction caused by the production of these items." Other income streams that we juggle include guided tours of our homestead, podcasts and YouTube videos in which Mikey and I share skills that we have learned, and this year a book that I wrote called *The Good Life Lab: Radical Experiments in Hands-On Living* (Storey). "Another income stream," I added with a shrug. "Assuming it continues to sell well, that is." I boasted that *Publisher's Weekly* just named my book best summer read and crossed my fingers. None of these things alone could keep us afloat, but collectively we do just fine. By having a diversified income plan we are able to turn on and off different income streams when we need to and this gives us freedom. That summer, we planned to turn off our eBay store and disappear with our backpacks into the wilderness.

"As much as we can, we avoid using money," I shared as a valued tip. We treat money as the low road, especially in exchanges with friends who make things. Once you make something you have something to barter, trade, and gift, and this is worth more than money. Barter comes second to gifting and is used in exchanges with acquaintances. Money is saved for total strangers and represents the *absence of any relationship.*

Marisa asked me about the mistakes we have made. I rolled my eyes dramatically, wondering where to begin. "I'll tell you the single biggest mistake." She nodded. I told her how early on, tempted by the lure of money and unsure about how we'd support ourselves, we accepted freelance work. Immediately we found ourselves back at the pump buying petroleum and standing in line at the markets buying expensive food and consumer goods. Freelance work took us away from making high-quality goods and had us buying inferior versions. It replaced the joy we had known from being makers of things with worry and stress. Most of the goods we had to buy came at a cost to the life of the world in the form of pollution, sweat labor, and fuel used in distribution, not to mention the lives wasted sitting at desks, dogs stuck at home in apartments, and sunsets missed. Freelance work derailed us from our own creative projects by replacing them with someone else's priorities. Once voluntarily unemployed and out in New Mexico, Mikey had time to tinker and learn to make things for pleasure alone. He made a fire trampoline, a watering system that turned on based on the dryness of the soil, and a low-power LED light system for indoor growing. These projects were immediately shelved once he took work for money. Likewise the garden (Figure 1-2) wilted from not being tended, my sewing projects were halted, and I found myself paying someone else to do repairs around the house. "We saw the problem right away, Marisa," I said with a harrumph. "We never made that mistake again." Today we firmly tell people who try to hire us that we are not for sale.

**Figure 1-2.** *Wendy examines her winter garden. Photo credit: Mikey Sklar.*

Mikey and I have chosen time over money. If you think about it, most people who choose money use it to buy time. We also reunited leisure and labor. Good labor is leisure—if you know how to look at it. We wrote a credo:

*When the whole world is for sale, it is a revolutionary act to become a maker of things.*

This is how our lives feel to us: meaningful and at times revolutionary. By following the pledges we made at the start of our adventure, we have become abundant rather than wealthy. We are no longer subject to the rise and fall of markets. The investments we have today take the form of tools; "buy things that make other things" is our motto. Regardless of the value of money, we can make the things that we need in order to live. From where we are standing today, having a job would be expensive. For example, a job would prevent us from discovering that we enjoy trail running out in nature and long-distance backpacking in the wilderness. I wouldn't have written a book or learned to garden. We are planning a trip to Italy to run the Alps with nothing but backpacks. We have the time.

I told Marisa about the tourists we took hiking days earlier. "They thought it was odd that we spontaneously invited them to hike on a Monday morning." "Is it

Monday?" Mikey asked as though dialing up an imaginary calendar. "We don't have many reasons to know the day of the week."

**Figure 1-3.** *Wendy and Mikey's car runs on biodiesel, available in your local dumpster. Photo credit: Mikey Sklar.*

That same night, from a hot spring in our yard, we watched a lunar eclipse that lasted from midnight to around 3 a.m. Marisa leaned over to me and confessed, "My family are billionaires who run big corporations. What they do comes at great cost to life."

"Everyone alive today is the problem and the solution," I told her, hoping to provide relief. I shared a theory I have that people land in one of two groups: some think life is full of abundance, and others think there is not enough. The former share, need little, and don't hoard. The latter suffer from a modern form of poverty, having enough and not knowing it. "Now, how do we get those relatives of yours to realize that they have enough?" I asked with a smile and an elbow. I wondered if Marisa is a sign of a change in the tide. Leaving her family's wealth must have been difficult. Though Marisa's life experience is very different from my own, we agreed that the discovery of freedom will likely take as many forms as there are people alive to reclaim it. "Dream aloud!" I shouted, as I locked the gate behind her. She was heading to Arizona to meet another couple that made a break for it and quit their jobs to live in a trailer. The sun was low on the horizon. It was late for starting anything. I cozied up on a bed in my garden that I made by welding together a rusty antique frame I had found in the trash. I covered it in soft foam and layers of linen. In summer Mikey and I sleep on it under a blanket of stars. We enjoy not running an air conditioner. I remember that everything I love is free.

Wendy Jehanara Tremayne is interested in creating a decommodified life. She was a creative director in a marketing firm in New York City before moving to Truth or Consequences, New Mexico, where she built an off-the-grid homestead with her partner, Mikey Sklar. She is founder of the non-profit, textile repurposing event Swap-O-Rama-Rama, which is celebrated in over 100 cities around the world, a conceptual artist, event producer, yogi, gardener, ultra runner, backpacker, and writer. She has written for *Craft*'s webzine, *Make:*, and *Sufi* magazines and, with Mikey Sklar, keeps the blog *Holy Scrap*. She is author of *The Good Life Lab: Radical Experiments in Hands-On Living* (Storey), a memoir and tutorials. *Publisher's Weekly* named it best summer read for 2013 and the book was awarded the 2014 Nautilus Book Silver Award for Green Living/Sustainability. Photo credit: Mikey Sklar.

# 2

# Interview: Tindie's Emile Petrone

---

WRITTEN BY **ADAM WOLF**

---

**Adam Wolf: You created Tindie (*http://tindie.com*), a marketplace for "indie electronics." It's often described as Etsy for Makers. How would you describe it?**

Emile Petrone: The way that we describe Tindie is a marketplace for the things you love. Basically, it started off as a site for my personal hobbies—DIY electronics, Arduino, and all that fun stuff, and I couldn't find a good marketplace for what I was interested in. It started out purely for my personal hobbies, turned into a hobby of itself, and obviously then a business. But, the way we look at it is basically just a marketplace for whatever you love. Whether that's Raspberry Pi, Arduino, robots, Daft Punk, or who knows what—we're fine with that. We're not so much focused on indie electronics or Makers or anything in particular; we take a much bigger picture view of what people can do with the platform.

**Adam: Is Tindie still tied to things that the seller has created, or are you open to resellers?**

Emile: There's two answers to that question. One is: "Does the site tailor to people creating what they're selling?" I think the answer is definitely yes. One thing we had to figure out was how to handle people who wanted to sell bulk components or supplies. Basically, once we rolled out Tindie Markets, the idea was that those people could build their own space where people can buy supplies or components and put all those products over there. Tindie is what you make of it—you can follow whatever markets you're interested in, and that's what we'll show you on the site. If you're interested in components, then you can find them, but the vast majority of people come to Tindie for interesting projects they can't find anywhere else.

**Adam: Let's back up a bit. Tindie is currently your full-time gig, right?**

Emile: That's right. Before Tindie, I basically had been around Silicon Valley and startups for five to seven years. I'm a self-taught engineer. My last job was being a web engineer for another startup that acquired the previous startup I had been at. I had been in sales, transitioned to web engineering, and was doing that full time.

**Adam: In our world, I see a lot of engineers who become self-taught sales-folk, where they have to learn sales after they've built their product. Were there advantages to approaching this area from the other side—from sales, transitioning into engineering?**

Emile: I think that if you're ever going to start a company, whether it's a web startup or a hardware company, or ever be creating and selling a physical product, at the end of the day, the ability to both create and sell whatever you're doing is essential. Whether that's your personal skill, or you and your cofounder, or however you distribute the skill set, you have to have that core. If that means you have to learn it on your own, you're going to have to learn it. That's what I realized doing sales [in Silicon Valley], if I was ever going to build my own startup, then I was going to have to learn how to code. The idea that I could hire or find a cofounder who could build whatever my vision was was not realistic. Five years ago, you'd hear that a lot. People would say, "I can outsource the engineering" or "my cofounder can handle all that" with that person having no experience in technical hiring, not understanding what were the technical skills they were looking for, and having no understanding of the process. That's my perspective coming from sales to engineering. I took a year off, lived off my savings, did the ramen thing for a year, while going through Python tutorials and learning Django. "I've got a year to do this, and it's either I learn it or I don't. This is the opportunity because I'm not married and I don't have kids."

Coming at it from an engineering perspective, if that means you need to sell, you don't have to take a year off to learn how to sell, but you're going to have to understand and master that craft if you think your company is going to go anywhere. You're going to have to sell either yourself, or the business, or the service, or the product to somebody.

**Adam: Through Tindie, you've been involved in many, many Maker-style projects. Hundreds, or even thousands of projects are on Tindie. Kickstarter and other crowdfunding options are a popular option for folks looking to get some capital to start a manufacturing run. Based on what you've seen, what do you think about crowdfunding Maker products?**

Emile: As of July 2014, we've got somewhere north of 2,500 products on our site.

**Adam: Wow!**

Emile: In terms of crowdfunding, I think it boils down to one question—have you ever shipped your physical products before? I think that we're looking at a time when crowdfunding platforms clearly have value. I'm not saying that they don't, but they're of questionable value to the smaller subset of people who use the platform to ship physical products, because 90-something percent of them have never actually designed, manufactured, and shipped physical products in volumes of one or two, let alone hundreds of thousands. This is their first time, and then they say, "I've got a great idea. I've been able to prototype it, it looks nice and neat, and we have a good enclosure. It looks all shiny and ready to go. Now we'll go to crowdfunding and get it out the door." The reality is those people still haven't designed and shipped something they've made themselves and they're doing themselves a disservice because they don't accurately understand the process. They can't accurately price their product. They don't know the expenses that will go into it, and the customer also suffers because they're depending on someone who has no understanding of what the actual process is that they're about to go through in their campaign.

The advice I give—I think it's the right way to do it—is that if you haven't ever shipped actual product, you need to basically start doing that. The easiest way to do that is small. Maybe a bare PCB of your design, maybe a kit. A real example: Nomiku (*http://www.nomiku.com/*), the sous vide Kickstarter, they started out by just building kits in their house. Friends and family would buy those and [Nomiku] refined the design, got it tailored to what it needed to be, and at the same time learned about sourcing and manufacturing, and basically grew their skills at business at the same rate. As you iron that out, you can look at assembly and more assembled products and the means to outsource the whole thing, where someone else manufactures the finished product, someone else handles fulfillment and shipping. After you've gone through enough iterations that you think the product is ready for the mass market, then I think it's totally fine and acceptable to get crowdfunding because you've gone through the process multiple times, and you've iterated as the project has grown, locating suppliers, finding manufacturers, finding ones at scale that are actually realistic for the expectations behind your crowdfunding project, and then it's basically a question of how much demand is there, and turning up the volume to fit that demand. You've already got the manufacturing process, you've already got those relationships. It's all ironed out with a real pricing and you know what your costs are. Now you can actually go and do a campaign correctly and smoothly!

I think that's the right way to approach it, but everyone likes to go a million miles an hour and everyone thinks they're smarter than everyone else. Until the end of time, we're going to have people who think this project is ready and go into crowdfunding, and it's bad for them and it's bad for customers. I am pro-crowdfunding, for the people who have a real understanding and the experience to back that up.

**Adam: I totally agree. One of the things I warn people about who ask me about crowdfunding their idea is that they are locking themselves into the product they show on this video, before they know what the best possible version of that product is. If they don't know what that product is, it can really damage the product and the company.**

Yeah, your reputation will be forever tied to this campaign that collected however many dollars and never fulfilled its promises. It's a double-edged sword—be sure you're swinging in the right direction!

**Adam: I'd like to ask a few questions about Tindie itself. Tindie has investors, right?**

Yup, we do.

**Adam: Did you have any difficulty explaining the Maker Movement or open source hardware to investors? Do you have any tips for readers choosing to look for investors for their new companies?**

Emile: I didn't have any difficulty explaining the Maker Movement or open hardware. The answer is no, for the simple reason that the investors that invested were the ones who already believed that something was going on—there's a change that's happening. They'd already convinced themselves that there's something there, and it was a matter of what was Tindie's growth like, and is Tindie a viable solution for what's out there? Will people start to coalesce around Tindie?

In the first three months, I still had my day job, and Tindie was my nights and weekends. Tindie was doubling in size month after month. That was just how it worked—there wasn't any "growth hacking." I was in the right place, at the right time. Investors who invested early on, they saw that and jumped at that point. That was not very difficult. Once we raised the rest of our funding last fall—that's a different story. The space was starting to mature, things were starting to happen, so it was a matter of showing the growth over the last year and where we thought it'd go—we think it'll be big, but we're not sure how big. We had some history of success, and we were still the leading marketplace for what's happening. That was a little different, because we had a longer history where we could show the growth we've had.

I think that for a business that's going the investor route, if you're going to look for outside investment, you have to have an exit for the investor. The two likely outcomes for that are either an acquisition or you go public and have an IPO. I think for the vast majority of hardware projects out there—they're not in that camp and they don't fit that model. They won't be a million-dollar business, and they aren't looking to create a strong presence in the space. A lot of projects are too small for that, but that's the realistic perspective venture capital is coming from. Smaller investors, angels, friends, family might have different expectations, and that's more the traditional "investor for a small business" relationship. I don't have the best view into that—I come from the startup, Silicon Valley world where the idea is to build a billion-dollar company. That's the perspective people have to come at this in order to make VC an appropriate tool to be used to build your business.

**Adam: Tindie obviously handles hosting a website, letting customers select and pay for products, and notifying the seller that they've sold a product, but what sorts of things does Tindie do that might not be obvious at first glance?**

The reason why Tindie works is scale and community. That is our core. When someone comes to Tindie and says, "I want to sell," or "I want to buy," well, we've got customers now in 82 countries around the world. That goes from hobbyists to the biggest organizations in the world. NASA, Google, Apple, Intel, government agencies, military academies, MIT, Harvard, Media Lab, everyone in terms of scale and scope. That's the community you're joining and trying to sell to. As far as additional features, the key part then is that we take on the cost of development and maintenance and improvements. You don't have to do that yourself. Let's say you were able to build a community online. Heartbleed, the OpenSSL vulnerability, is something we handle at Tindie. If you were running your own site, you have to manage and stay abreast of security concerns. We're constantly being looked at for fraud and by hackers. With Tindie, you don't have to worry about that. We want to take on as much of the process of actually selling your design and your products as possible, so you can go on doing what you love, which is most likely creating new things, new projects. That's the main service we provide for sellers, along with a large community that's diverse. We're always coming up with new designs to increase conversions, new tools to make your job easier, and new ways of selling and getting your products in front of people who are interested in what you're doing. I think that any business can do one thing really well, and ours is to make a marketplace. If you're a small business, chances are the one thing you do well is innovative new products. Running a website where people can buy stuff, that isn't it. Let people take care of those things so you can do what you do best.

**Adam: I'm about ready to wrap this up—is there anything cool and new at Tindie that just came out or will come out soon that you want to talk about?**

Emile: Our long-term plan now is that Markets will continue to expand and get more and more diverse. It came from a realization that because we're a marketplace and because we depend upon people listing products on our site, there's no reasonable way for us to manage the taxonomy and categories on the site when we have to react every day. By giving that to the community, and saying, "You have full control over what categories get made. Make a market for your passion!" I think that naturally we'll expand into interesting areas that we didn't foresee. As an example, yesterday a Daft Punk market (*https://www.tindie.com/m/DaftPunk/*) was created for people that are creating Daft Punk pieces. I think that'll be exciting to see as that grows and matures. We'll see helmets and crazy different Daft Punk pieces on the site. We also have markets around hackerspaces. There's a tymkrs (*https://www.tindie.com/m/tymkrs/*) marketplace now. There's different platform marketplaces now, like Spark (*https://www.tindie.com/m/sparkcore/*) and Electric Imp (*https://www.tindie.com/m/ElectricImp/*). We didn't see that coming, and now we see that it's great. As the weeks and months progress, that's only going to get bigger and bigger, and more and more interesting, and let Tindie become a platform for whatever you're interested in. Tindarians can join the markets that they're interested in, and Tindie becomes a site that shows them only products and projects that they're interested in. If you don't care about supplies, you won't see supplies. If you're really into Star Wars and Daft Punk, for example, then we'll have tons of stuff that rallies around your interests.

We don't have any real new features coming out except to support that. It's been about a week since we've had Markets, and we've had 45 or 50 created already. It'll be very exciting to see where that evolves.

**Adam: Excellent! Nice talking to you, Emile!**

Emile Petrone is founder and CEO of Tindie (*http://tindie.com*), the marketplace for innovation. After seeing so many interesting projects pop up over the last few years, he built the marketplace as a way for Makers to bring their creations to market. In a year and a half, more than 2,000 products have been listed by over 400 inventors. Photo credit: Emile Petrone.

Adam Wolf is a cofounder and engineer at Wayne and Layne, LLC, where he designs kits and interactive exhibits. He also does embedded systems work at an engineering design services firm in Minneapolis, Minnesota. When he isn't making things blink or talk to each other, he's spending time with his wife and son. Photo credit: Adam Wolf.

# 3

# Evolve With the Maker Scene

WRITTEN BY **ALEX DYBA**

My passion for Making and electronics started as a kid in Russia. My father was an engineer and told me, "Never ask someone to do what you can do yourself." We had a soldering iron and would work on making scale models of trucks and tanks in our spare time. I watched and helped, but mostly, when no one was around, I would plug in the soldering iron and melt stuff. The culture of making ran in the family. My grandfather was a tile worker who was often sent to far regions of the country to decorate apartments of bigwig politicians. My grandmother used to make enough canned goods from her garden plot (called a "dacha") to last all winter. As kids, we ran around looking for parts for our bikes, fixing our inner tubes by cutting up and gluing together other inner tubes.

Around that time we got our first computer, a TRS-80 clone that could load BASIC programs from cassette tapes! Dad and I later modified it by adding a joystick port to play Lode Runner. That opened my eyes to the idea that there is a way to make technology do what you want. Although when I was little, I was not successful at making things do anything that worked properly; my strengths were in taking things apart, with the intention of later putting them back together. Video games were also great but unfortunately they were all in English so that forced me to learn a second language. Looking back, there were many times when `Format C:\` seemed like the correct command to run on the 286 if I wanted to play a game. Legos and all kinds of construction sets were also around when I grew up, needless to say.

**Figure 3-1.** *The GetLoFi store at Maker Faire in San Mateo, California. Photo credit: Alex Dyba.*

Right before the Soviet Union's collapse, we left Moscow for the United States and settled in the small farm town of New London, Iowa, where I spend a year getting a feel for rural life while practicing English, playing Super Nintendo, watching Beavis and Butthead, and enjoying American sports: a crash course in Americanization if there was one. This experience was not all devoid of continuing electronic and Making explorations. Luckily, the basement of our house was completely empty so I could set up a little lab. Another bonus was that the school library knew my of interest in electronics because I checked out every book on the subject. When a donation of vintage radio and TV repair books came in, they offered them to me. We didn't have very much money, but my mom did buy me the RadioShack 200 electronic set, and boy, was that awesome! I was finally able to successfully make electronic circuits that worked properly.

In the late '90s and most of the aughts, my life was a pretty ordinary American experience. I lived in the second-largest city in Iowa. Home to Rockwell Collins Radio, Cedar Rapids still plays a very important part in the aerospace industry, from the invention of the two-way radio to communications for space missions. Though I was never officially part of the ham radio community, I did benefit from going to a swap meet and learning about the Rockwell surplus store, open once a week and brimful of decommissioned equipment and industrial prototypes.

# 6699

Having gone through the process of becoming a Maker of a thing, you should not have to relearn all the steps. Try your best to document—if not for yourself, then for others who might want to undertake this same journey. Maybe by sharing, you'll discover a solution that only exists in the minds of other people on the Internet.

The Internet wasn't available to me until freshman year in high school, and it was pretty much a done deal that I would use it in some way to make a living. Reading through some old essays of mine, the idea of selling things online was firmly planted in my brain at an early age. Familiar with the fact that "one man's trash is another man's treasure," I began selling surplus electronics on eBay. That was always a fun time, but unfortunately contributed to a hoarding habit of keeping pieces of equipment with the intent of eventually selling them for a profit.

Another pivotal point in my Making career was discovery of *circuit bending*. This involves modifying children's toys through short-circuiting to make new sounds. Little information was available at the time on the Internet about how these instruments were made, and *Make:* magazine didn't exist...heck, Google was considered second fiddle to Yahoo, Lycos, and Ask Jeeves. Dogpile was the search engine of choice because it aggregated all other search results into two pages. Blogger was still its own project for anyone interested in publishing a website with RSS feeds, a way of sending your message to the world. I wanted to reach out and document my experiments in circuit bending, so I started *http://GetLoFi.com*.

At first it was thrilling to know that someone out there on the Internet—mind you, this was before Facebook and even MySpace—was interested in what I was doing, or—gasp!—doing the same thing. Growing up in the second-largest city in Iowa did not exactly bring together like-minded people in the field of circuit bending, but people across the globe bonded and found each other through GetLoFi much as everyone does now on Facebook. With similar interests in modifying musical instruments, strangers began to contact me and the circuit-bending scene picked up steam, driven by the newfound novelty of the Internet. Artists and creators came out of their basements and onto the Web, showing more projects, more secrets, and ultimately organizing events such as the Bent Festival, an in-person gathering where lifelong friends were made.

GetLoFi's mission was to demystify the world of circuit bending and create a curated community resource. My failures and successes were often blogged to the world, and many people connected through the website, which is still active but less instantaneous: updates are now made on a Facebook news feed. In addition to sharing my projects, I wrote posts demonstrating ways to cut corners or to do things correctly. Eventually, one person asked me for a copy of my project: the classic Atari Punk Console circuit, designed on a printed circuit board. The nature of the online community is to share, so I ordered all the parts, etched a few circuit boards, put the two together in baggies with instructions, and my kit business was born!

The biggest fear to overcome was the feeling of unveiling something you worked hard on to someone who may not understand it. How will they react? It's kind of like the fear of public speaking or performing music, but once you get out there it's on and there is no turning back.

The most well-known GetLoFi product came out much later and was probably the biggest hit and disappointment at the same time. The Nintendo Guitar (Figure 3-2) was the first commercially available six-string instrument that used an authentic Nintendo Entertainment System console as the body with a salvaged guitar neck. About 30 were made and sold, with many people still asking me to create more. It was innovative, nostalgic, and fun. It got on some TV shows and musicians used them in their videos, but commercially making them at a reasonable price in a reasonable time was a total disaster. The guitars were 100% handmade instruments and often took hours of precision labor to correctly assemble. The worst part was selling out of product and taking backorders. With deadlines looming, it got harder and harder to enjoy putting together my creation, when really it should be fun! That experience changed my mentality about what handmade really means and how much of a living can you make doing that. Time is valuable and given all the extraneous duties of the professional Maker, the costs can creep out of the price range of people who want them. There is a fine line between what steps should be automated and what parts should be artisan crafted.

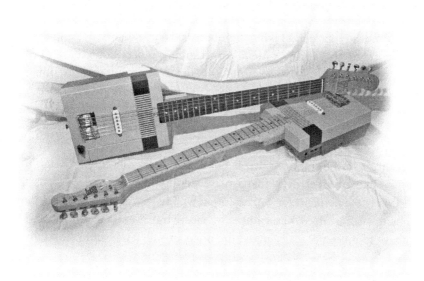

**Figure 3-2.** *The NES Guitar used a discarded game console as the body. Photo credit: Alex Dyba.*

Besides having successful or challenging products, establishing relationships with your customers, clients, and collaborators is probably the best part of being a Maker Pro. Every sale is unique, and both sides have an interest in the transaction, potentially leading to future sales. Think of going to a big-name fast-food restaurant and ordering something: the cashier takes your money and gives you what you ordered—that's it! The person working the cash register honestly doesn't care if what you got was good or not; there is no need to discuss or new innovative ways to use that sandwich, you just eat it. You exchange pleasantries, part ways, and no one remembers that transaction unless something went horribly wrong. Now think about a craft fair or Internet website, where the person ordering knows who you are, and may have seen your products before, or heard about them from someone else. As a seller, you are thrilled to have made the sale (which will keep your dream afloat another day), and in return you honestly care whether your customer likes the product or not, because your future sales depend on it. There will be bad transactions, and mistakes happen, but that is life. The best we all can do is learn from our experiences, good and bad. In the end, for me personally, it is a pleasure to know that in the future someone will open up a strange little wooden box and discover a circuit board with GetLoFi clearly written on it, or a guitar with my signature on the inside.

When it comes to repeated Making, do not trust your memory. Having gone through the process of becoming a Maker of a thing, you should not have to relearn all the steps. Try your best to document—if not for yourself, then for others who might want to undertake this same journey. Maybe by sharing, you'll discover a solution that only exists in the minds of other people on the Internet. Instructables, *Make:* magazine, iFixit, and YouTube have made empires hosting user-generated knowledge.

Being a commercial full-time Maker is not an easy task. All your costs start to add up even if you have a relatively inexpensive solution to a problem. It's possible that eventually the demand for your solution will disappear because all of the current problems will be fixed, or a competitor steers traffic to their similar product, so you have to innovate, scale, adjust the price, or walk away and just do this as a hobby.

So why do people choose to be a professional Maker? There is a saying, "Pick your battles." Obviously, if you or I begin designing a microcontroller platform as a challenger to Arduino with similar specs, we would not succeed. The market share that Arduino already possesses is substantial. Why? Arduino has been around for a very long time in Internet years. There is an established user base, name recognition, and whole slew of compatible third-party accessories. The price point is low. There is brick-and-mortar representation at RadioShack, Micro Center, and Fry's. Everybody who is a Maker or electronics enthusiast knows of it. Arduino clones from China are so cheap that parts comprising the circuit board are more expensive than the finished product, shipped! At this point it's fair to say that what we have is about as good as it's gonna get. It's in the bull's-eye of price versus features. Why are products like Raspberry Pi, Galileo, and BeagleBone successful? Well, because they do things that Arduino cannot do. They have features that Makers who designed them wanted and thought that other Makers would want them. If you want it, someone else will want it too.

The tricky part is wanting things that more and more people will want as well —a need that has no easy, inexpensive solution. It's always a battle of time versus money. However, most people with money value their time, so they are willing to pay extra for a product that works right out of the box.

Consider that when making things for others. Ultimately your skilled labor and knowhow is more economical to someone who, to accomplish the same task, would first have to study and learn about the problem, find a solution to the problem, acquire all the tools and materials, and set aside time to come up with a finished solution. At that point the "person with a problem" becomes a Maker! For some people, reading schematics or programming a computer is magic, yet these people can be the most inspirational and creative because they have a specific idea they wish to make a reality. When they use your product to think outside the box, that can inspire a future product that would cater to their niche and perhaps improve on the initial application.

Commercial Maker products can be unique and human. Consider one of my current projects, the Bottle Cap Knobs (*http://bottlecapknobs.com*) shown in Figure 3-3. These are simple replacement knobs for guitar or amplifiers, inspired by the Cigar Box Guitar aesthetic of reusing trash in the theme of drinking, singing, and smoking. Crafting with discarded materials has long been a pastime. Some examples include trench art during World War II, where intricate carvings were made on spent artillery shells. Bottle caps in particular have been used in numerous projects. DeKalb-based Maker Austin "Paw Paw" Cliffe makes beautiful contact microphones from plastic and metal caps, and his work inspired mine.

By incorporating a previously discarded item, each knob turns out to be unique in a certain way and also has sentimental value to me as a Maker. I remember drinking some of those beers and sodas! I started saving caps from everywhere I go: San Francisco, Iowa, Chicago, Louisville, and even Russia. Non-Maker friends also contribute by stashing their bottle caps, after being taught the proper way to pry them off without bending the design part.

Transforming bottle caps into a usable item is tricky. The back side of caps is usually covered in plastic sealer to prevent gas from escaping and also to prevent rust. This surface does not glue well and the tops are metal, so punching holes through them is hard and ruins the design. Thankfully, the perfect solution ended up being the precision of a laser cutter. I cut reclaimed wooden planks to the exact size of the bottle lip, and then use a little homemade compression jig and a bottle capping bell to crimp the caps back, just like on the original bottle! The caps also fit perfectly on knurled shaft potentiometers and virtually anyone can install them to customize their piece of gear, in a sense Making a new Maker who is not afraid to stand out and be different. They may even be inspired by the fact that all the materials are free, and the clever use of them makes the final product unique. These caps sell for about a dollar: it is not much, but is nearly 100% profit in the end, minus a little bit of personal time, the usual primary investment of every entrepreneur or professional Maker.

**Figure 3-3.** *Bottle cap knobs add a personal touch to potentiometers. Photo credit: Alex Dyba.*

Just like soul food which tastes better when the cook puts their heart into it, products made by Makers feel special because an individual's time, which they can never get back, is at core of the object that they share with you.

Alex Dyba was born in Moscow, Russia, while it was still the former U.S.S.R. and currently resides in Minneapolis, Minnesota. His hobbies include playing music as Talking Computron, constantly inventing new products, and drinking whiskey with friends and family. His favorite football team is the Vikings. Photo credit: Alex Dyba.

# 4

# Making It

WRITTEN BY **JIMMY DIRESTA**

As a very young Maker, I would see things that I wanted and immediately begin to figure out how to make them. At about the age of 9 or 10, I began to make dart-shooting crossbows. The trigger mechanism was a challenge: I would make a trigger and test it, and then another one until I perfected it. The challenge was to puncture an old car tire or a basketball. Most kids want to play basketball; I developed a dart gun that would puncture the skin of the ball. Making a mechanism like this requires thousands of tiny decisions and trial and error: the spring tension in the trigger, the strength of the projectile bow or rubber band, the balance of the darts. As a kid, I was not aware of the lessons I was learning. I learned things that I still think of and use daily. Making a mechanical hand was another thing I wanted to make as a kid. I loved figuring out the various ways to mimic the human hand, exploring the way joints would close and spring open. Eventually I graduated to gas motors and motorcycles and then cars, antique machinery, and tools...curiosity and obsessiveness are the most important things in the early development of a professional Maker.

What makes one a professional? I get this question often. I have been a student and teacher at the NYC School of Visual Arts (SVA) for over 20 years, and I've had the opportunity to watch many young people grow up and become "professionals." There are several factors I observe and practice in pursuit of professionalism.

Early on, I would take any job to develop my experience. I learned back then that people with a prior experience in any given job were likely to get hired. I worked at a sign shop in high school, starting as a helper. At home I would cut letters on my band saw. One day I proved I was good enough to cut letters on the sign shop's band saw. Eventually I would sit most of the day, cutting all the letters for all the signs. I was 18, and the next best-skilled guy was 50. Some 30 years later, I use a

CNC machine to cut most of my signage. Putting the time in to become skilled is the most important thing. Like learning to play a guitar, the path to perfection is practice.

**❝❞**

Most kids want to play basketball; I developed a dart gun that would puncture the skin of the ball.

After my college education was over, I took any pay I was given. I was gaining experience and developing a foundation to one day be able to raise my day rate... and keep raising that number. I still take jobs for free to learn new things. I do this to keep my problem-solving skills sharp and in tune. When I first purchased my CNC router, I needed to learn on real jobs. I made anything my clients needed, simply to use and learn the software. Staying humble and looking for the lesson in any situation always helps me.

**Figure 4-1.** *Jimmy DiResta in his workshop. Photo credit: Stett Holbrook.*

I don't like to hear people complain. I never want to sound like that. Whether it was long hours, laborious work, or bad bosses, I never complained. I would simply add to my list of what I want and what I don't want. Separate the type of people I would not like to work for from the good people for whom I would work for free! The good people leave you feeling proud of what you do and you always learn from them.

I spent many years going to China to have toys manufactured. I was the engineer and designer. I was the problem solver. I worked mostly freelance. The most valuable thing I learned from working with the Chinese is to be self sufficient. The factories and agents I worked with would handle all aspects of the project. I "handle it" when doing work for my clients; I like to handle all aspects of the job. I always like to be able to do it all myself in the event I can't hire a subcontractor for a given task. This also gives me an education on all the different parts of any job, so I can help communicate with a subcontractor and understand their job.

When working with the factories and engineers in China, I would hear the phrase *momentai* all the time. This means "no problem." Like MacGyver, it is important to improvise and figure it out. A close friend always says to me, "make it go right!" I hear her in my mind in times of trouble. I make it a personal challenge to figure something out and exhaust all possibilities. I play a game with myself to go into a situation and figure it all out, there and then, with any resources available.

About 15 years ago, I was on a job installing shelves. I first thought the wall was Sheetrock, but I was wrong: the wall was brick. Without the proper screws, I thought I would have to return to install another day. However, I then had a brainstorm: why not drill the holes in the brick and hammer a piece of my wood pencil into the hole, and this gave me a spot to screw in a typical wood screw. (Luckily, I had about eight pencils in my bag.) Ever since then, I always keep wooden pegs with me for any brick wall installation. It works better than most of the solutions you can buy. You always learn something new about yourself in situations like these.

It is the simple things that will distinguish you as a professional. Always be on time, early if possible. Always have pen, paper, and a ruler with you! A big pet peeve of mine is when I meet with a "designer" and they take my pen! This is a person who can't think ahead. (When they want my pen *and* ruler, it is a bad sign.) Always be prepared for anything at all meetings. Bring pictures, and take pictures. No fumbling! Be precise.

Being honest is a must with your clients. If you begin to lie and make excuses to your clients, you'll look foolish in the long term. I always take responsibility. Clean up any miscommunication; it can lead to expectations that are not met. It's important to try to get it all in the emails and communications recorded. Your clients will respect you if you have all your work in order. Above all, always deliver on time and on budget.

Always have an answer. Never say "I don't know"—it is important to always have the toolbox in your mind open and ready to solve problems, or at least begin dialog to solve a problem for your clients. Dig into your experience and come up with an answer! Never be at a loss for words. I often challenge my students to improvise an answer to a given question. For example, Q: why did you chose this color? A: I don't know. Wrong! A: I chose this color because it complements the color of your shoes! Anything is better than "I don't know." I recently saw a TED talk where the speaker told viewers to "fake it 'til you make it" and this is exactly what I am talking about here. If you begin a dialog in a design meeting by trying to avoid looking unprepared, you just might come up with the solution!

**Figure 4-2.** *DiResta's workshop is packed with tools, materials, and old projects. Photo credit: Jimmy DiResta.*

When working on TV with my brother, he would always say "never say no to an improvement," and in design it is the same way. It could lead you somewhere you never thought you could be.

## Making Guitars and Toys

One of my last teachers at SVA was toy inventor Mark Setteducati (*http://en.wikipe dia.org/wiki/Mark_Setteducati*); he encouraged me to be an inventor. Right after school, I began to help him develop toy ideas and magic tricks. He introduced me to several people in the toy business and my work expanded, all the while exercising and improving my invention and problem-solving abilities. At school I was experimenting with making guitars. So I was doing this at the same time I was developing inventions.

A friend introduced me to a guitar shop and I made a few guitars for them, One of the owners knew guitarist Steve Vai (*http://en.wikipedia.org/wiki/Steve_Vai*) of Frank Zappa and David Lee Roth band fame.

So I made my first celebrity guitar for Steve Vai. He played it on David Letterman the day I gave it to him! It was 1990, I was out of school for five months, and I felt like a pro! The very next day my girlfriend was working at a store as the register girl. There was a rock-and-roll guy that would flirt with her. One day she told him about me and that I made guitars. She showed him the photo of the Steve Vai guitar. He had seen it the night before on Letterman and immediately wanted to meet me. He name was Adam Holland, a professional guitarist with an endorsement deal with ESP guitars. So I began to make him a few guitars and developed a relationship with ESP and added a few nice pieces to my portfolio.

In the spring of 2013, my friend and former accountant was working with Wyclef Jean (*http://en.wikipedia.org/wiki/Wyclef_Jean*), raising money for his new tour. Wyclef mentioned he wanted a signature guitar in the shape of a gun. He wanted to illustrate the concept of turning guns into guitars, in response to street violence in Chicago. My friend introduced us, and that April 13 we came up with the gold AK-47 (*https://www.youtube.com/watch?v=6XgmjHrTJGo*) guitar, a fully-functional electric guitar built off of an assault rifle frame (Figure 4-3).

**Figure 4-3.** *DiResta made this AK-47 guitar for musician Wyclef Jean. Photo credit: Jimmy DiResta.*

In addition to guitars, I also built toys. From the moment I left school in 1990 to about 2008, I developed toys and mechanical devices related to toys. I engineered and invented several toys in the craft category and a few gross-out toys. My biggest invention was a called Gurglin' Gutz (*http://en.wikipedia.org/wiki/Gur gling_guts*), a squishy ball shaped as a heart, a brain, and an eyeball. It started a category of squishy toys and balls that continue to get more gross with each passing season.

I still do toy invention and consulting. My most recent project was to help the Basic Fun toy company reintroduce the View-Master. We redeveloped the images for the new licenses and the plastic housing. I was fully immersed in the world of stereoscopic images, and the process of how they are developed and created was a great learning experience of a great product with a rich history.

## Television

One day my brother John asked me if I could film him picking through the trash and making a table from what he found. At the time, he was an actor between TV jobs in Hollywood and would make things from wood he found in the trash and sell them at the local flea market. It was 2002, and I had been experimenting with video editing with Final Cut, so I went out to record him and put together a seven-minute video. His agent was pleasantly surprised that the tape and edit were actually good. Because of the tape, we got a meeting at FX Networks, and I put together a presentation of ideas for a full TV season. I was hoping to get a job on the show as a behind-the-scenes designer/producer, so I had my portfolio and a big book of ideas at the meeting: all hand sketched and photoshopped images of various ideas for episodes. The producer we met with at FX asked if I wanted to be on camera. This was very unexpected; I was just hoping for a job on set at the very least. I said yes, of course, and the producer liked the idea of me being the "designer" brother and John being the "funny" brother.

A few months later we were taping a show named "Trash to Cash." We shot seven episodes. The show was all about me and John making things from the trash; it was my first TV show.

I went on to make a show called "Hammered" (the working title was "Making It with John and Jimmy"), we made the pitch tape for this show on our own, without a network or production company. My friend was a hairdresser; one of her clients was a network executive at HGTV. She told the executive about me. I sent her a tape, and 12 months later the show "Hammered" was being made for HGTV. It aired 28 episodes in 2006-2007.

In the summer of 2004, I made a pitch tape for a concept called "Lord of the Fleas." It was about finding trash and reworking it for sale at a flea market. Everyone loved the tape, but no TV deal was offered, so I just put it on my YouTube channel. It was getting some hits, and a few people loved the idea but it was not getting any traction. In 2010, a friend asked me to visit him at the new production company he worked at. They needed new signage, and I was going to make it for them. While there, I asked if they took submissions for ideas, and sent my friend the YouTube link for "Lord of the Fleas." The owner of the production company called me that evening to ask if he could show it around. He showed it to the Discovery Channel that week, and 10 months later we were in production. Discovery aired the show with a new title, "Dirty Money." In the summer of 2011, we did one season that is available on Netflix. When it seemed like we were not getting a second season due to a in-house shakeup at the executive level inside of the Discovery network, I decided to keep in touch with my audience on YouTube. I posted my first video in fall of 2011, and since then I have posted over 125 videos of me making things in fast forward. I now have fans all over the world and the beautiful emails and comments inspire me to make more!

**Figure 4-4.** *DiResta and his brother John pose by a doghouse they built for "Hammered."*
*Photo credit: Jimmy DiResta.*

The journey is half the fun, and you never know where your next job will come from. It is important to take all inquiries seriously. You just never know. The few things I mentioned early on in this article are the things I practice daily. If it makes me a professional, I guess I am. I never forget the things that I learn along the way that can make the journey more fulfilling. Technical and/or personal: it all helps make you a pro!

Jimmy DiResta is an experienced television personality as an on-camera designer/builder host. Jimmy combines his charisma, his innate ability to solve problems with found objects, and a wide variety of tools and techniques, with sweat and brute force. He has appeared on the Discovery Channel, HGTV, DIY, and FX Networks as a host and co-host. Photo credit: Stett Holbrook.

# 5

# The Power of Constraints

WRITTEN BY **MICHAEL KRUMPUS**

If someone had approached me five years ago and told me that today I'd be running a business selling electronics that I designed, I would have told them they were absolutely crazy. My response might have been something like, "Surely you must have me confused with someone else. I'm a software engineer and don't know anything about electronics. I have a solid, rewarding career and there's no way I'd be brave enough to make things and sell them. I just don't do that kind of thing. You've got the wrong guy."

Yet here I am, running an electronics company called nootropic design (*https://nootropicdesign.com*), selling products that I designed myself to thousands of customers throughout the world. And I'm doing it from my basement while still working full time at my day job. Granted, it's not very far-fetched for someone to learn some new skills and start a small business. There are thousands of young entrepreneurial spirits out there creating new technologies, running crowdfunding campaigns to get started, and living on caffeine and ramen. But I started when I was 40, and in a different field than I had spent my career. It has been a completely unexpected twist in my life, and the most rewarding thing I've ever done.

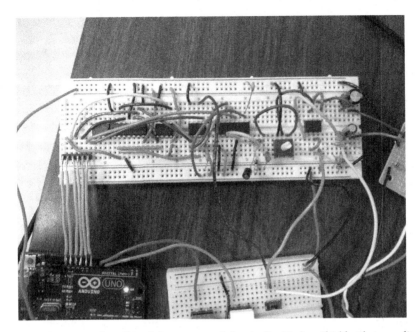

**Figure 5-1.** *A breadboarded prototype of the Audio Hacker Shield. Photo credit: Michael Krumpus.*

It all started with an Arduino, the ubiquitous electronics prototyping board at the heart of a million hobbyist projects, art installations, robots, and, indeed, and entire cottage industry of add-on products. I ordered an Arduino in 2009 and started with everyone's first program, blinking an LED. For all you programmers, blinking an LED is the hardware equivalent of writing a "Hello, world" program when you are learning a new language. It started there for me just like everyone else, and then came the obsession. I started learning everything I could about what I could do with this little Arduino board. I connected other components like LEDs and motors, and learned as much about electronics as I could. I learned from the work of others, I read data sheets for components (after learning what data sheets *were*), and spent late nights in the basement wiring up circuits and occasionally creating a bit of smoke. Everywhere I looked I saw electronics, and I was now beginning to understand how it all worked. The appliances in the kitchen, the dashboard of my car, everywhere. It was as if a fog had been lifted. I spent more and more time in the basement tinkering.

But that's not like me. I don't do that sort of thing. I had been in the technology world my whole career, so where was all this new passion coming from? Why was this so different? In time, the answer became clear...

## The Power of Constraints

As a software developer, I quickly found that writing code for a microcontroller like the one in the Arduino presented me with a very constrained environment. There is only 2K of RAM to work with (that's kilobytes, not megabytes!) and the speed is only 16 MHz. No video, no audio, and just a handful of I/O pins to interface with other components. You might think that this would be discouraging, but the truth is quite the opposite. Designers of all types have long known that constraints are what unleash creativity. Constraints present you with constant challenge. Constraints force you to find clever ways to accomplish your goal. They make you invent new techniques, find shortcuts, evaluate trade-offs, and in my case, squeeze every bit of capability out of a tiny microcontroller. During my years of learning, I've figured out how to do pretty heavy stuff with an Arduino, like processing video signals, making action arcade games, and digital signal processing with audio.

One day I noticed a spike in web traffic on my site from one particular location: the Department of Homeland Security. Hundreds of distinct visitors from Homeland Security throughout the Washington, D.C. area were looking at the Defusable Clock. I was sweating bullets all day, thinking that someone was going to bust down the door.

I've played around with more powerful developer boards like the Raspberry Pi which runs Linux, is more than 40 times faster than the Arduino, and has 262,000 times as much memory. Sorry, but this doesn't really excite me. I know people have done some great things with these more powerful technologies, but it's not as much fun for me. I'm naturally frugal, so I relish doing more with less.

Think about it this way: if someone presented you with the most powerful hardware in the world, with all the tools and materials you could imagine, and told you that you could design and build anything you wanted, what would you do? I mean, you can do anything, right? This might sound good on paper, but with no constraints, the creative mind doesn't really engage. I'm sure some of you could make good use of this situation, but as for me, I need some serious boundaries. That's what sparks my creativity, and I've heard the same from dozens of other technology designers. So embrace constraints and know that it is the secret sauce you might need to unleash your creativity.

## The Power of New Capabilities

With my passion and creativity ignited by the constraints of tiny, simple electronics, I started having ideas. Not just projects, but products. The Arduino has launched a vibrant industry of add-on boards, called "shields" (like the prototype in Figure 5-2) as well as other devices that have the Arduino microcontroller at their core. Using the Arduino taught lots of people how to program microcontrollers, so naturally many people have designed add-on shields and all sorts of other gizmos and brought them to market. Motor controllers, robotics platforms, GPS receivers, network interfaces, music players, and gadgets with colorful, blinky LEDs. LOTS of LEDs.

**Figure 5-2.** *A prototype of nootropic designs' Video Experimenter Shield. Photo credit: Michael Krumpus.*

But could I do that? I had the ideas and some technical skills to design electronics (just barely), but could I actually make something I could sell? A hundred questions raced through my mind day and night. How would I design a circuit board? How would I get it fabricated? How would I source the other parts? Where would I sell my goods? Should I set up my own ecommerce site and sell directly to customers? What about shipping? How do you process credit cards? Would I even make any money?

I felt like I was trying to solve a giant puzzle. One by one, I figured out how to do everything required to sell my own product. I learned circuit board layout. I found a printed circuit board manufacturer to make the boards for a reasonable price, as well as options for small batch prototyping. I scoured the planet for the best prices on components. I set up an ecommerce website to accept credit card payments and learned the pitfalls of web hosting providers. I set up a blog and support forums to help my customers. I learned how to ship packages anywhere in the world and how to print postage labels at home. I learned how to pack merchandise carefully and where to get the best shipping supplies.

I gained these capabilities by finding the right tools and services that provide them. I think of this toolset as a platform on which I operate. There are lots of choices for each of these things, so an aspiring Maker Pro has a lot of research to do. The great news is that you can build a platform of capabilities with free and cheap tools that are only getting better. Having an operating platform is the difference between simply having a dream and being able to execute on it.

It was hard to set all this up. It took time. It was frustrating and I had problems, but it's worth the investment. Some of these capabilities like having circuit boards manufactured in small batches, or sourcing components in small batches, simply did not exist 10-15 years ago. These new capabilities and the plethora of choices for your platform of tools are at the heart of the Maker Movement. These new capabilities give dreamers and creators the ability to execute. As they say, ideas are a dime a dozen. Lots of people have ideas, but few do anything about it. Creativity and inspiration make ideas, but execution makes things happen.

## Failure

No discussion of entrepreneurship would be complete without mentioning the importance of failure. This often comes in the form of a perky twenty-something startup founder telling us how great it is to fail so we can learn from our mistakes. "It's OK to fail! In fact, it's awesome to fail!" he says with a smile on his face.

I kinda hate that guy.

Grudgingly, I admit he has a point. I have learned a lot from my mistakes and missteps, as painful as they were. Failure sucks, but I have learned from it. You can't make me like it, and I'm not going to smile about it. Building technology and running a business are full of surprises and unexpected twists, many of them unpleasant.

My first product was a real flop. It was a simple Arduino shield that allowed people to have more output pins so they could do more. I was sure it would be a big hit! Unfortunately, since I was so new to electronics, I had no idea that there were already products on the market and they were much better and cheaper than mine. I didn't know they existed because I didn't even know the right search terms to use when looking for competing products on the Internet (I should have searched for "I/O expander"). In other words, I was such a newbie that I didn't know the right terminology for the product I was presumably "inventing"! Naturally, I overestimated the demand and spent money and time getting ready for a deluge of orders that never came. I was discouraged when business didn't materialize, but curiosity and drive kept me going. Since I had no business, I had plenty of time to start my next product. How's that for optimism? Well, I had no choice.

## Maturing as a Maker Pro

Despite a slow start, four years later I have eight products and a few of them are very successful. My most successful product is the Defusable Clock (Figure 5-3). It is an electronics kit that lets you build an alarm clock that looks like a bomb. I sell the electronics and you put it on materials that look like explosives. Sticks of phony dynamite, whatever. My customers have built some amazing devices, and I keep a gallery of submissions on my site.

This is clearly a provocative product and it has gotten a lot of attention in magazines, newspapers, and countless websites. A great way to get attention is to push people's buttons. Try to hit a nerve. And take great photos of your product that people want to publish!

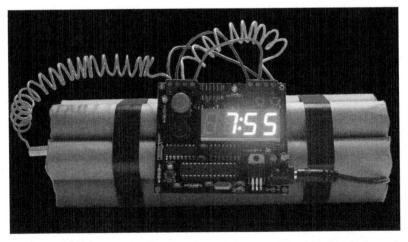

**Figure 5-3.** *Michael Krumpus' Defusable Clock is no more dangerous than any other clock. Photo credit: Michael Krumpus.*

Not all attention is welcome. One day I noticed (using Google Analytics) a spike in web traffic on my site from one particular location: the Department of Homeland Security. Hundreds of distinct visitors from Homeland Security throughout the Washington D.C. area were looking at the Defusable Clock. I was sweating bullets all day, thinking that someone was going to bust down the door. And I was scheduled to fly to New York City the following week to attend Maker Faire. Would I be on the no-fly list? In the end, nothing came of it, and years later I'm still selling the Defusable Clock, sometimes to law enforcement agencies for bomb squad training!

Obviously I've learned a lot about running a business: taxes, supporting customers, managing a supply chain, and all the other things you'd expect. More importantly, I've learned a lot about myself. I now know that I truly love technology for the joy of discovery that it brings. I do not aspire to move ever higher in a company's management ranks, because that would just move me further from real technology. I've learned courage to try new things (and fail), and to be more comfortable in the spotlight despite being an introvert. I've learned to be thick-skinned to negative feedback and resilient to unexpected problems. I've also recognized that most surprises are actually positive, not negative.

One more story about the Defusable Clock. My website has the following plea for understanding:

> This is only a clock. If you are with the FBI, Department of Homeland Security, CIA, ATF, Department of Defense, National Counterterrorism Center, Interpol, or SEAL Team Six, please know that I am on your side! So, we're cool, right?

I am currently working with the U.S. Navy to provide Defusable Clocks for them to use in training exercises. Ironically, it's entirely possible that SEAL Team Six may actually use it for training. Now that's an unexpected twist I can smile about.

Michael Krumpus is a software engineer and hardware designer in Minneapolis, Minnesota. He has a master's degree in computer science from the University of Minnesota and has 25 years of experience in software design. Photo credit: Michael Krumpus.

# 6

# Have Makerspaces Made the Traditional Artist Studio Obsolete?

Written by **Susan Solarz**

Buckminster Fuller often said the true job of people is to return to whatever it was they were thinking about before somebody came along and told them they had to earn a living. For me, it was the visual arts. In my early twenties, I studied studio arts and supplemented my income from cocktail-waitressing by trading my drawings and paintings for groceries, haircuts, and meals. However, once I tried working in three dimensions, I knew that sculpture was my passion.

Despite that, parents and advisors warned that the art world was too competitive and pushed me to choose another career. Because I like being outside, I chose biology. That bad decision launched a series of events that led from college to 6 years of graduate school to 10 years of hopping from job to job, either rejecting long-term offers (and wondering why) or failing to thrive or even fit in. In 2012 I was suffering in a particularly stifling job and realized I had to make some changes. In January of 2013, I recommitted myself to working as a sculptor.

I was unsuccessful in my search for affordable studio space until I discovered Twin Cities Maker, an organization housed in a mid-century light-industrial building in the Seward neighborhood of Minneapolis. My multimedia work requires both space and access to expensive equipment such as a laser cutter and welding tools. At TC Maker's hackerspace, the Hack Factory, I found everything I needed and began to explore my creative ideas and develop my artistic vision.

The Hack Factory is better suited for many visual artists than traditional studio rentals. It is spacious enough that as I prepared a large installation, piece by piece, I arranged the entire system at the space in order to make edits. As a rule, a project that members are actively building may be stored on site until completion —find a studio that can offer that and still cost under $100/month!

**❝❞**

There is a strong community of support within the organization. It feels like other members want to see me succeed almost as much as I do. At the Hack Factory, I found the supportive and encouraging environment that I have always wanted.

Members seem to be interested in everyone's projects and are very generous with their time—I have received hundreds of hours of volunteer help on my sculptures from more than a dozen other members.

**Figure 6-1.** *Susan Solarz reclines in her Origami Chair. Photo copyright E. Katie Holm, http://katieholm.com.*

Hackerspace members are people who enjoy making things. So, one might assume that members volunteer to help in order to learn a new technique or out of a self-interested curiosity. This does not seem to be the case. In addition to "Make. Learn. Teach," the TC Maker slogan, there is another tenet: "Be excellent to each other," that is important to members. There is a strong community of support within the organization. It feels like other members want to see me succeed almost as much as I do. At the Hack Factory, I found the supportive and encouraging environment that I have always wanted.

There is always another member who is willing to learn about my latest obstacle and brainstorm with me. Sometimes a group of members will share ideas and even work together to help me identify one or two options worth trying. If that doesn't work, then staff members, who already volunteer many hours to keep the facility and machines in top shape, are incredibly generous with their time. They often help members through difficult procedures.

The organization provides access to all the equipment that is common to wood and metal shops. My membership grants me the freedom to use almost any material, including fiberglass, acrylic, and metals. I rely on our shared equipment for creating my artwork. If I could not build for myself the components of my sculptures that require machines like the metal-band saw, table saws, drill presses, and the Bridgeport mill, then my sculptures would be too expensive for me to make.

Working artists need more than a place to create their artwork, however. We also need facilities for meeting with prospective clients and collaborators. The Hack Factory offers three conference rooms and encourages members to use these spaces for collaboration and meetings. There is also a large classroom and all members are encouraged to share their expertise by teaching a class. Instructors determine the price of the class per student and keep half of the total amount—another boon for artists who struggle to make ends meet. I will teach my first class this summer on using fiberglass—a skill that I taught myself last year at the Hack Factory.

## Origami Rocker

My first project was to redesign a Frank Lloyd Wright chair. The original Wright Origami Chair famously has a tendency to tip forward, especially when the sitter scoots forward. Instead of revising his design, Wright "solved" the problem by adding anti-tipping feet (with metal tips to make them look intentional). Years earlier, I tried to redesign the chair with limited success and wanted to try again.

Using Hack Factory equipment, I made the front feet larger. However, the chair still had a tipping problem. The user wasn't in any danger, but the experience of getting up from the chair could still be startling—an undesirable quality to be sure!

**Figure 6-2.** *Susan Solarz's Origami Rocker chair improves on a classic. Photo credit: Michael Troemel.*

As I dragged my failed plywood model to the "up for grabs" woodpile, another hackerspace member mentioned that a chair with a tipping problem might make a great rocker. I was too frustrated to consider it at the time, but the next day I realized that a rocker was the perfect solution. This was the first of many times when ideas from other Makers helped me to overcome obstacles.

It took some research to learn what the arc of the rockers should be. It turns out that the arc for any rocking chair can be drawn on a circle with a diameter equal to seat height x Pi. After I determined the arc shape, I carved a model rocker out of polystyrene foam and hired an expert to bend two square tubes to match the model. I then made several steel plates to screw to the feet and tail of the chair. The next step was to weld some narrow tubes to the plates and then weld the rockers onto the tubes. However, I did not know how to weld!

Fortunately, the Hack Factory offers a variety of classes, including welding. After my weekend course in MIG and TIG welding, I was good enough to attach the rockers to the chair. I then carved an ergonomic seat for the chair from spray foam covered in body filler. Then it was time to test the seat for comfort.

Minne-Faire, a two-day event highlighting creations from the local Maker community, was approaching, so I decided to display the unfinished chair in order to gain feedback. My carving mistake became clear after the first several sitters: every woman who tried the seat loved it, and every man found the back half of the center ridge too high! I adjusted the shape and confirmed that the adjustment was enough to make the seat comfortable for both men and women.

For a sporty, modern look I decided to give the chair a high-gloss fiberglass finish. No one at the Hack Factory had experience with fiberglass so I learned how to work with it by watching YouTube videos.

Most of the chair received one layer of fiberglass and four to five layers of resin. I thought that tinting the fiberglass resin with universal pigments would eliminate the need to paint the chair. However, I soon realized that resin is extremely translucent because it contains no opaque base. Therefore, the two different colors of body filler I used to create the contoured seat were very visible, even through three layers of fiberglass cloth and resin.

The translucence and surface problems meant I had to paint the chair. After pricing spray guns, paint, and the rest of the necessary equipment, I decided to save money and have my chair painted white at an auto body shop. I also sprayed white rubber onto the bottom and sides of the rockers, similar to a pair of Chuck Taylor shoes, as a sporty-looking way to protect floors from scratches. I am looking forward to my first formal showing of the Origami Chair this fall.

## Musical PVC Bench

In the summer of 2013, I received a small grant to create a temporary exhibit in downtown Saint Paul. The city has limited outdoor seating, which often forces pedestrians to use architectural features of buildings to rest. To draw attention to the need for outdoor seating, I made a bench that would both function well outdoors and demonstrate that public seating need not be boring.

I settled on creating an S-shaped "musical bench" out of vertically oriented PVC pipes. The pipes are very sturdy and spray paint would protect them from yellowing in the sun. Using pipes too short to touch the ground allows rain to flow straight through the bench, making it usable almost immediately after a rain or snow melt. The pipes that do not touch the ground are also free to resonate when struck at the top with a foam paddle. I made the bench interactive and "musical" by providing foam paddles for users. One can sit and rest, or pick up a paddle and make musical notes or play a short tune.

**Figure 6-3.** *Susan Solarz and PVC musical bench. Photo by Jeffrey Thompson / Minnesota Public Radio News. Copyright 2013, Minnesota Public Radio. Used with permission. All rights reserved.*

I created the design using SketchUp, researched solvent welding as a means to bond the pipe segments, and bought the necessary materials. A fellow Hack Factory member pointed out that there is almost no surface area at the point of contact of two circles so I increased the surface area where the pipes would be joined in order to strengthen the bond. After struggling with various saws, another insightful Maker created a jig so I could shave off the same amount of PVC with each pass of a router.

Even with the jig, I was worried about finishing the bench on time. With less than 3 weeks before the bench was to be displayed, I had 270 PVC pipe pieces to cut, shave, and weld. I also needed to sand the finished bench top, apply spray paint, make foam paddles, and attach the paddles to the bench with wire rope. Again, my Maker community provided enormous assistance.

I used the TC Maker blog and social media to outline my challenge and recruit volunteers. Six members altogether volunteered about 100 hours of their time, making it possible for me to finish the bench on schedule. In September, the Musical PVC Bench was publicly displayed and received enthusiastic praise from the public and even coverage in some local media outlets. The bench has since been donated to the Dayton's Bluff community as a contribution to their neighborhood improvement project.

## Conclusion

Makerspaces are better suited for making art than a traditional studio rental. The supportive and encouraging environment, as well as the expertise of other members, has been absolutely critical to me in realizing my projects. This environment could be of benefit to other artists as well, though to-date few other working artists have signed up. However, with easy access to equipment and meeting rooms, helpful classes, the option to store active projects and the low cost, I believe many other artists would thrive in the makerspace community.

Susan Solarz is a sculptor who focuses on simple furniture for homes, businesses, and public spaces. She is drawn to projects that offer an opportunity to problem-solve, such as designing an outdoor bench that is dry almost immediately after a rain, or a single piece of furniture that can serve as either side table or stool. Whenever she can, she up-cycles scrap industrial materials such as steel conduit and PVC. All of her work encourages human interaction in a safe and comfortable environment. Photo credit: Michael Troemel.

# 7

# Pick Your Maker Sherpas

WRITTEN BY **ROB KLINGBERG**

Several weeks into my newfound obsession with tiny lighting, after I'd already decided to embark on the journey of turning my rough prototypes into commercially viable products, my wife asked a very good question: "If the need for a product like yours is so great, how come somebody else hasn't already made it?" I've held that question in my head through the ups and downs of the last three years, and in its answer lies the difference between startup success and failure.

Makers today sit at the pinnacle of decades of advances in electronics, manufacturing, engineering, and staffing—all of which enable us to birth something new and launch it into an eager marketplace with the lowest barriers to entry in history. Twenty years ago, we'd have needed factories. We'd have needed minimum order quantities requiring tens of thousands of dollars in up-front investment. We'd likely have had to quit our day jobs. And we'd either have had to go back to school to learn the skills needed to launch our product, or pay to hire someone who had. Today, all we need is a PayPal account, access to the Internet, and a sometimes-borderline crazy belief that our idea is awesome.

All of this is seductive, maybe too seductive. It becomes easy to think we can do this thing, this work of bringing our idea to market (and then raking in the dough that will inevitably follow), without anyone's help. I've discovered quite the opposite to be true, and that's the story I want to tell here.

Since 2011, I've run Brickstuff, a company that manufactures and sells micro-lighting and automation products (seen in Figures 7-1 through 7-3) primarily for the Adult Fans of Lego (AFOL) community worldwide. Customers use our modular system to add lighting and effects to Lego models they build. While our sales are by no means astounding, they have doubled each of the last two years, and we shipped product to customers in 22 countries in 2013. My title is "Chief Enthusiast," which is really just a fancy name for the guy who designs, manufactures, packages, sells, and markets everything. I'm the one up late with the soldering iron making battery packs for starter kits, I'm the one driving six hours in the snow to an AFOL convention, I'm the one tracking down a lost shipment sent to the Philippines, and I'm the one hunched over data sheets and trying to master the ever-bewildering *http://digikey.com* search interface.

**❝❞**

Electronics design firms wanted $10,000 to engage and a minimum of one million units. (What is it with that number, anyway?)

**Figure 7-1.** *Brickstuff makes tiny lights that fit inside Lego bricks. Photo credit: Rob Klingberg.*

I also have three young kids, a wife who works full time, and a full-time day job. I still don't know if this business will evolve to the point where it could sustain my family (earning me the much sought-after privilege of being able to pull the ripcord and parachute out of the Day Job jumbo jet), but there is one thing about which I am dead certain: I wouldn't be where I am today without the passion, sweat, labor, and ideas of an army of people I'll never meet. These people have been my teachers, my designers, my naysayers, my manufacturers, my retail representatives, my buzz-generating team, my support structure. They've taught me the difference between crimp-style and insulation-displacement connectors, they showed me how to cook circuit boards in a toaster oven, they've made sure I appreciated the importance of paying as much attention to the packaging as I did to what's inside, and they've shown me that I, too, can master international postal shipping regulations. In the end, these lessons have made all the difference, but in the beginning, almost everyone had the same question: "Lights? For Lego? Why would you want to make that?"

The answer to that question requires a bit of explanation. I was an electronics tinkerer as a child, disassembling every toy, gift, and model train I received (with any luck, I was also able to put things back together). I also dabbled in programming, but due to frustration with math in high school, my passion for technology entered a hibernation period and I went on to major in English in college. A string of communications, writing, and sales jobs followed, but my tinkerer's mind was always alive just below the surface. While my career progressed, huge changes were occurring in multiple industries, reducing barriers to entry for Makers and democratizing technology to an unprecedented extent. I didn't fully appreciate the extent of the change until two unrelated developments converged on my dining room table. The first development came in late 2010 in the form of a slender, black microcontroller called Arduino. I forget where I first learned about the board and its capabilities, but immediately I knew this was a platform even an English major could get his hands around (Processing, on which Arduino was based, was originally developed for artists). I placed my first order from Sparkfun (order #298718, for two Arduino Pro Minis) on New Year's Eve, 2010. I, like so many Makers, owe a tremendous debt of gratitude to Massimo Banzi and everyone associated with the Arduino project. The electrical engineers in the room never like to hear this part of the story, but it was the Arduino that enabled an English major like me to build a viable business on microcontroller programs that brought tiny LEDs to life. If I'd have been stuck learning assembly, things never would have gotten off the ground.

The second development came into my house and onto my dining room table in the form of a big, expensive box of Lego bricks: the Grand Emporium kit (#10211 for the AFOLs reading this). This was my first large-scale Lego kit, and after building it, I stared in amazement at the level of detail present in the model, especially the light elements. This was not the Lego of my youth, where the color and shape palettes were small; this was a serious modeling medium capable of reproducing almost any detail. The Grand Emporium had small lanterns outside the main entrance, a streetlight built into the sidewalk, and a gorgeous chandelier hanging from the top floor. Beautiful though these constructions were, they were unlit. It was at that point that Lego and Arduino came together, and I knew what I had to build. Lego would be the canvas, and Arduino-animated SMD LEDs would be the paint. All of which brings me back to my wife's very good question: "If the need for a product like yours is so great, how come somebody else hasn't already made it?"

I would soon have the answer to my wife's question. While several other people had started Lego lighting businesses, none offered products that achieved the polish and simplicity I believed AFOL's were seeking in a lighting product. The need seemed large (due to Lego's explosive popularity worldwide), but how to begin? The barrier to entry was low: I had my domain name, Twitter handle, workshop in the basement of my house, and several orders of parts from DigiKey and Mouser on my doorstep in no time. This was the seductive part: it all seemed so easy to get the parts I needed, and thanks to Arduino, I had the platform necessary to bring my ideas to life. Why couldn't I do all of this myself? How hard could it be? I would soon learn just how little I could actually accomplish without help.

As any Maker who has worked to commercialize an idea knows, getting this far in the product ideation process is the manufacturing equivalent of making it to Everest Base Camp: you've come farther than most people, but the hard part really hasn't begun. You can see the summit, but it's much farther away than you imagine. You must approach the task with the right mix of bravado and humility, and you'd better not try to reach the summit alone.

I knew from the beginning that I wanted to build Brickstuff so it was able to grow to theoretically limitless scale if needed, but at the same time I needed to start small enough that I could afford to bring product to market using a self-funded model without going broke. In my experience, this is the toughest part of starting up: starting as small as you can but building everything (process, product, suppliers, prices) so it can scale to high volumes if the idea takes off. I found it to be like trying to balance a marble on a sheet of glass: lots of small adjustments and constant motion are needed to keep things from coming off the rails. Lots of people drop their marbles; not everyone reaches the summit of Everest. I wanted to be the guy balancing his marble while climbing.

It didn't take long before I stumbled in my ascent, and I quickly learned why more people hadn't done what I was trying to do. After building my first prototype, I turned to what I believed were the traditional avenues to get it produced:

manufacturing design firms, electronics production firms, wire and cable companies, and marketing firms.

I'll never forget my first meeting with a large manufacturing design firm. I drove to their offices in downtown Minneapolis on a blistering summer day with my Lego Grand Emporium model and makeshift lights in the trunk of my car. I carried everything up the stairs to their beautiful offices, and set up the demo unit on their large conference table. Engineers and marketers came into the room, each handing me a business card as they entered. I had worked hard to get this meeting —only through a friend was I able to secure time with the firm's top people.

After demonstrating everything and outlining my business plan, the killer question came: "So, is it fair to say you have less than one million dollars to invest in this project?" Because apparently that's what people with good ideas needed: $1 million just to get started. No doubt they deduced from the blood draining into my feet that perhaps they'd aimed too high with their initial salvo, and they kindly offered instead to conduct a component lifecycle assessment for a mere $14,000.

At that point, it was fair to say I was rather nauseous. Is this really what it was going to take? Visions of desperate Kickstarter campaigns complete with hastily-made pitch videos flashed through my head, followed by desperate calls to friends and family to get them to invest in my crazy scheme. I packed up my things, thanked everyone for their time, politely declined both their offers, and drove home.

For several more weeks, it was the same story with the other traditional professionals I tried to engage. Electronics design firms wanted $10,000 to engage and a minimum of one million units. (What is it with that number, anyway?) LED lighting companies weren't any better, when they bothered to respond to my inquiries at all. And I paid a local marketing company $3,500 to develop a corporate identity with logos I'm fairly certain my 12-year-old could have topped with a couple hours on Adobe Illustrator.

**Figure 7-2.** *Brickstuff's lighting controllers feature several different lighting patterns. Photo credit: Rob Klingberg.*

Don't get me wrong: each of the companies I tried to engage had its place. Each had been in business for a decade or more, and each appeared to be going strong. The problem wasn't them; it was me. They were offering helicopter rides off the top of Everest, but I'd just left Base Camp. I soon learned it was possible to cast a monumental amount of money into a bottomless pit and receive very little in return that was in any way useful for my stage of development. I didn't make much forward progress until I stopped looking in the traditional places and decided to see if I could get the democratization of just about every industry to work for me instead of against me. The right approach was right in front of me, though initially I couldn't see it. I describe it to people as a feeling similar to that felt by Harry Potter when he first saw Diagon Alley: a hidden world, with everything you need, hiding in plain sight. All I needed to do was turn my head a bit. My Diagon Alley was *http://elance.com*, one of the more popular freelance job sites that are transforming the way work happens. Just one example: while still slugging it out with the local marketing firm over the lackluster results of their $3,500 corporate identity project, on a whim I posted the same project request to Elance. Within 60 minutes, I had responses from all over the world, most at price points that were a fraction of my local provider. After researching the portfolios of the responding companies, I selected a firm in Argentina and sent $150 via PayPal. A week later, I had just what I wanted: clever and unique logos and logotypes that continue to define my company today. I negotiated a 50% refund from my local firm, and set about thinking what else I could send over to the Elance collective.

Over the next several months, I had many more positive experiences on Elance. I hired a company in Malaysia and received one of the best market research reports I've ever seen for $100; this helped me refine my European marketing strategy and was invaluable in helping me identify key influencers in the AFOL community. I hired a designer in Boston for $185 and received trade-show banners that rival Fortune 500 companies in their quality. Most significantly, I hired an electronics engineer in Australia who continues to be the mastermind behind all of my circuit board designs. Indirectly through connections first made on Elance, I also found one of the best wire and cable manufacturers in the world. Having my Elance contractor vouch for me allowed me to place initial orders at half the supplier's typical volume (being able to order 500 of something is better than having to order 1,000 before you even know if it's going to sell).

When U.S.-based suppliers wouldn't even return my initial inquiry emails, even the largest Chinese and Taiwanese firms replied within hours, treating me with the professionalism my email signature seemed to require. If only they knew the truth: I was a one-man operation in a converted basement coal room in St. Paul, Minnesota!

Once my aspirations seemed back on track and once again semi-viable, I scoured data sheets to identify prominent LED manufacturers globally, then sent a blind request to one of the largest firms in China (the same company that manufactures products for Ikea). Once again, my email was promptly returned, along with custom specifications and a price quote. Once I saw the manufacturing cost per item, I knew the business models I'd built were viable. One month after sending funds via PayPal to a street address in Shenzhen that didn't even resolve on Google Maps, 2,000 hand-made LED light strips, complete with my corporate website address printed beneath custom crystal-clear epoxy, appeared on my doorstep. I wondered what the fancy Minneapolis engineering firm would have charged me to get even half this far.

**Figure 7-3.** *Want to light up a whole block of Lego buildings? Just use more controllers. Photo credit: Rob Klingberg.*

Truth be told, there were some negative experiences as well—there always are when you take the solo-sourcing path. Several Elance contractors failed to live up to their promises, but I attribute this more to good intentions of breaking into the freelance market while holding down a day job than to anything malicious. Thanks to Elance's escrow process, the only commodity I lost was time. I'd still trade my negative experience with traditional firms for this "fail fast" experience any day.

Expanding beyond Elance, I've learned to enlist my customers as first-class citizens in my product planning. When I first launched my products for sale on the web, I had an elaborate array of individual items for sale, each of which combined (in my head, of course) in multiple ways to make the ideal lighting system. I posted my products and waited for the cash to roll in.

Two weeks later, I was still waiting. The problem? Nobody knew what to buy for their particular Lego creation. A prominent Lego blogger suggested I create several starter kits, and these continue to be bestsellers today. Learning to let go of some parts of the product development cycle has been a challenge (especially when each new product can require $3,000-$4,000 in R&D, design, and production costs), but in the end I'm in business so every one of my customers can have the same moment of "WOW" I first had with that Grand Emporium the day I lit up every one of the kit's model lamps.

Speaking with customers, and seeing the amazing things they create, continues to be one of the greatest joys of this entire journey. Waking up in the morning and seeing that a new order has come in overnight from Croatia is exhilarating (how does someone find out about Brickstuff in Croatia?); getting an email from a first-time customer saying they didn't have any idea how to add light to their Lego models until seeing my product is even more satisfying. My favorite experience, though, is when parents come to our table at a convention, see our product, and launch into dialog trying to convince their kids it would be a good idea to buy one of our starter kits—I love watching the parents position the purchase as good for their kids, when clearly I can see it is the parent who wants their child's permission to play like a kid again. These experiences and many more make all the late nights spent soldering yet another batch of battery packs more than worth it. Best of all, I know we will always be pushing the envelope; as industries continue to democratize and transform, I know there will always be armies of people I'll never meet who are ready to help me spot the cool ingredients and turn them into something that will make everyone's eyes light up.

In the end, maybe the best answer to my wife's question about why others hadn't tried to do the crazy things I was trying to do with Brickstuff is: because they weren't as lucky as I've been to have found such great sherpas to aid me in the climb to the summit.

When he's not working at his day job selling cloud software to large companies, Rob Klingberg moonlights as the Chief Enthusiast for Brickstuff, a company he founded in 2011 that manufactures lighting and automation products focused primarily on the Adult Fans of LEGO (AFOL) market worldwide. Rob grew up programming his Atari 800 in BASIC, but fell off the technology wagon in college, where he majored in English. He continues to be flattered by the number of people he meets through his work with Brickstuff who assume he is a formally trained electrical engineer—he won't tell if you don't. Rob lives with his wife and three children in Eagan, Minnesota. Photo credit: Rob Klingberg.

# 8

# I'm Not a Maker, I'm a Builder

WRITTEN BY **JOE MENO**

I am a Maker.

But I don't use 3D printers, or pliers, or hand tools. I use a toy. I use Lego bricks. I also use Lego wheels, beams, and plates. But I make more than Lego models. I make a lot more. It just so happens that the brick is only the beginning of the creations that I have made. Let me tell you the story of me and the brick, a story that is still being written.

The story begins decades ago when I was first introduced to Lego sets in Germany. I was an Army brat, and my family was stationed in Mainz, Germany. This was in the '70s, so in terms of entertainment, there was a movie theater that my dad managed and really, not much else. TV was only the black-and-white Armed Forces Network, and it was there that I first saw a show called *Star Trek*, which gave me a love of space and science. It was also here that I got the Moon Lander Lego set, which was my biggest introduction to the bricks. I began building and making my own things, mostly underwater stuff because I happened upon some Jacques Cousteau books in the library and became fascinated with the seas. Space also appealed to me, and I built moonships and even my own monorail (it had a turntable to reverse direction and looked very blocky). I kept on building until we came back to the States, and left much of my Lego collection behind.

By the time I got to junior high, I was pretty much a science geek. And while I wasn't much of a builder anymore, I did other things to push my creativity. I drew from books and played in the school band. For those who know about the Lego hobby, I had entered a dark age from building. Even then, I still had some parts left over at home in a Lego storage case. It was a visit years later when I was in college that I saw the case out to be taken away. I looked at the parts then, and made a simple tower on the table. My mom saw that, and kept the Lego bricks at home. At that time, there were new things that were being made by Lego, and I noticed. I would buy a set from time to time, including a pneumatics set, which had a little hand pump to operate a crane. How neat! I built the models in the instruction books and then put the set aside. At that time, I was working on a degree in design at North Carolina State's School of Design. I didn't really have time to play. And with a degree in hand, I took on the world and ended up in advertising. I didn't know it at the time, but I was slowly getting ready to come back to playing in a big way.

I built the set outside under the shade of a tree outside of Downtown Disney, and building it sparked something in me that I hadn't felt since college.

It took a few advertising jobs and more than a few years to realize that advertising as a career was not a field in which I wanted to spend my life. I did learn that I was able to create visual materials quickly if needed, but I lost interest when I was an art director at an agency that abruptly closed. Our main client left us, and the agency was dissolved a month later. I spent a few months living on unemployment and freelance assignments and then realizing that I was not getting any younger, decided to do something completely off the wall.

I went to Florida to work at Walt Disney World at the age of 34. For the previous four years, I had worked at the local Disney Store, just to keep social since I worked for a small agency. It was a simple transfer. I happened to like working at the store, so it wasn't much of a leap to go to the park. So I packed up my car and reported to DisneyQuest in Downtown Disney. I needed a second job to make ends meet. I applied to the Lego Imagination Center. These two places would create a completely unexpected foundation for the rest of my life.

My Lego job did not last very long, as I was given a position outside of the building watching the activity area for rowdy kids. I wilted in the Florida heat and humidity, but the one thing I got that made the job worth the effort was a discount at the store. It was 1998, and I bought the Star Wars X-Wing set that had just come out.

And I was hooked. I built the set outside under the shade of a tree outside of Downtown Disney, and building it sparked something in me that I hadn't felt since college. I bought some more sets and began to use my computer to go online to see if I could find any Lego items. What I found was sets for sale on eBay, and an online Lego community. I could buy other sets, and there were others like me! Suddenly my universe expanded to a worldwide network of builders. My Lego job didn't last all that long (six weeks, if you're wondering), but it was the job that pointed me to a horizon. After spending almost a year at Disney, I packed up and went back home to North Carolina, to a job working at a newspaper. This was 1999, and I was starting to build Lego stuff.

I discovered that Lego building was a fun hobby and soon got myself known online by showing my models on Brickshelf (*http://www.brickshelf.com/*), a gallery devoted to Lego builders. By 2001, I decided to go to a Lego fan convention, and I was surprised and delighted by the people I met there. All of them had a common thread in that they built stuff. They were Makers, just like me.

Except I also liked to explore. With the Lego fan community being so small at that time, it was a time of discovery for groups. Conventions were just starting to happen, and I found myself wanting to become more than a builder in this community. I wanted to make more than Lego models. I realized that my skill set in publishing and design was rare in the community, so I started creating a Lego community magazine that became the *BrickJournal* magazine (*http://www.brickjour nal.com/*). The first issue took the community—and even the Lego Group—by surprise with its high production values and articles. The first online issue was downloaded over 70,000 times and, by the third issue, downloads averaged 90,000.

I also began volunteering and went through the ranks in community events, eventually becoming an event coordinator for what was the biggest convention at that time, BrickFest 2006, in Tysons Corner, Virginia. From there, I stepped back for a year while working on the magazine—Lego was now interested in funding *BrickJournal*, and I found a publisher for it.

After nine issues and roughly two years, *BrickJournal* went from online to print in 2007. And with it, I became a professional magazine editor. My profile in the community grew because of my activities, and I helped to start conventions in 2008. And I was still exploring.

My Lego building was also getting known, with some models going viral after posting. I built the first Lego model of an iPad (Figure 8-1) when I wanted to know how one would feel in my hands—I just went to the Apple website to get specs and built a model to fit. The model was posted on Flickr and quickly went viral (Figure 8-2), as this model was built months before the first iPad was released for sale.

**Figure 8-1.** *BrickJournal taps into grownups and kids who love to build. Photo credit: Joe Meno.*

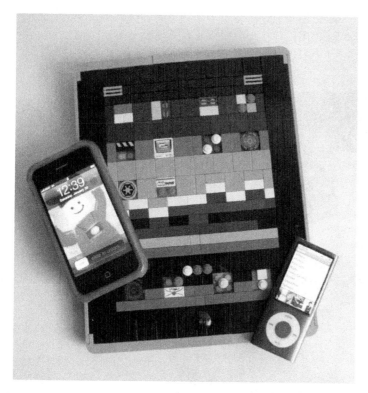

**Figure 8-2.** *Joe Meno built the Lego iPad to spec so it had the same heft as the real tablet. Photo credit: Joe Meno.*

Another model I built was of the robot Wall-E (Figure 8-3). It went viral because I finished him a couple of weeks before the movie was released. I try to keep building, if only to keep my skills honed and to keep my "legitimacy" as editor of *BrickJournal*.

**Figure 8-3.** *Joe Meno's Wall-E went viral before the movie even came out. Photo credit: Joe Meno.*

And I am still exploring. Because of the brick, I have built models that have been seen around the world and have been displayed in many places that I would have never would have thought of, like the National Air and Space Museum. I also have made friends here, there, and everywhere, including people who have worked in the space program and Disney animators! My latest model was of Olaf, the snowman from *Frozen* (Figure 8-4). My inspiration? One of the story artists at Disney who nudged me to start building the snowman!

**Figure 8-4.** *Joe Meno's Olaf model pays homage to the popular Disney movie Frozen. Photo credit: Joe Meno.*

I am also returning to childhood by volunteering at a local school for a budding FIRST Lego League program. While it is fun to explore, it's just as important to pass what you learn to the next generation. Hopefully, I will be able to give a small spark of inspiration to the kids as I teach them building and programming. There's so much more to build, and I am still going forward. I am making a magazine, models, and a career. All from a toy.

I'm not a Maker. I'm a builder.

Joe Meno is the founder and editor of *BrickJournal*, a website and bi-monthly magazine for AFOL. He has worked with the Lego Group on various projects, and is one of the coordinators of BrickMagic, a Lego fan event based in Raleigh, North Carolina. Photo credit: Joe Meno.

# 9

# Hacking Brick and Mortar

WRITTEN BY **ADAM WOLF**

One January day a few years ago, my phone rang. This was slightly unusual, but I recognized the number as one of our distributors, so I didn't let it go to voicemail.

"Hey, Adam! How fast could you ship an order of 10,000 Blinky POVs and 10,000 Blinky Grids?"

I've had phone calls that felt completely normal, and only in retrospect was it clear that they were life-altering. This was not one of those calls.

Matthew Beckler and I run a small electronics "kitbiz" called Wayne and Layne. The company first consisted of Matthew and myself, working nights and weekends, creating new kits for our customers to put together. At the time, we had a handful of through-hole kits, a variety of online distributors, and sold some kits directly. We shipped a few hundred kits a month, and I don't think we had ever had an invoice listing more than 250 of a single item. We had done a few design projects with clients, where we helped them flesh out an idea they had, or helped them make schematics and a PCB from a design they had prototyped. We spent about 10 hours a week on Wayne and Layne, sometimes more, sometimes less.

"Well, I'll have to check on a bunch of things, but if I absolutely had to guess, I'd say six to eight weeks--Chinese New Year is coming up, you know."

It was the end of January, and Chinese New Year was only few weeks away. The holiday, also known as the Spring Festival, is a big deal in many countries—like China, Hong Kong, Taiwan, and Malaysia, all countries that produce a lot of electronics. People get time off, and pretty much everything stops as folks get together with their families, clean house, eat a big meal, and set off fireworks. I'm certain most people look at it with fondness, but it strikes fear into the hearts of anyone who wrangles supply chains. If you deal with only one company, you'll usually be impacted by about 10 days—but not everyone has the same days off, so once you deal with more than one company, you can basically expect to lose February.

"Yup!" said our distributor, "we've been in talks with a large, international brick-and-mortar retailer. We're doing a bunch of new products with them, and we showed them your Blinky kits. They loved them!"

The Blinky POV and Blinky Grid (Figure 9-1) were two of the kits Matthew and I had developed. The Blinky Grid is a battery-powered 7 × 8 LED array with a microcontroller and two light sensors. It comes as parts, and after you put it together with a soldering iron, you can go to our website where there's a page with some JavaScript. The page lets you enter messages and a bit of artwork, and blinks two squares on the screen in black and white, back and forth. If you hold the light sensors to the squares, it wirelessly reprograms the stored information and updates your kit.

**Figure 9-1.** *Blinky Grid has been a successful project for Wayne and Layne.*

**❝❞**

We looked at our schedules, all our quotes, our contingency, and gave the go-ahead. I mean, if we screwed it up, it was only our entire business and more money than we had dealt with in our entire lives that we were risking.

The Blinky POV is almost the same, but has only one row of eight LEDs. To get the Blinky POV to display, you have to wave it through the air, where your brain reconstructs the rapid blinks of the LEDs into an image. This is the "persistence of vision" effect.

We developed the two Blinky kits after organizing some electronics and programming sessions at a summer camp for girls. With a different kit we had developed, we had so many difficulties using the school computers—driver installations needed administrator privileges and soldering issues confused the USB controllers—that we decided we needed a kit that could be updated and personalized via a computer without physically connecting to it, or installing software. A year of nights and weekends later, we had the Blinky Grid and Blinky POV.

"They're looking to order a sample order of around 10,000 POVs and 10,000 Grids. A few questions: can you do that, and if so, how long will it take, and how much will you charge?"

We discussed the opportunity for a few minutes. The distributor was very generous—they would handle the end supplier, some regulatory and legal issues, and they explicitly wanted us to keep our branding on the kits. I was super excited, but it was obvious that this was big, big enough that I had to bring Matthew into the loop before making any real decisions. I relayed the news to Matthew via IM. We decided our next step was to get quotes from our suppliers for a sufficient quantity of all the parts, with costs and lead times. Not all of our suppliers were really ready for an order of this size, so we also needed to shop around for alternatives. Of course, we needed to balance the risk of new suppliers not being able to deliver quality parts on time against existing suppliers not having capacity or being too expensive. We also had to decide if we actually wanted to go forward with this deal. We both thought it might be too large, and we didn't want to destroy Wayne and Layne.

Quality is an important part of every business, sure, but it's even more important when you're building open source hardware. When you encourage anyone to make copies of your product, business changes a little. It's almost like merging traditional engineering companies with the fashion industry. Any individual product you develop becomes less important, because anyone can copy it. Having a trusted seal of approval, a "brand," is one of the things that can distinguish you in the marketplace. The second half of that is to develop a good set of skills and processes so that you can continue to quickly create new products while supporting the existing base. At Wayne and Layne, we have a list of business and technological skills we want to develop, and it's an important factor when we are deciding which projects to work on.

We're heavily opinionated and we'd both like to end up working for Wayne and Layne full time, but we don't want to take outside investors, and we don't want to be acquired. Matthew and I are both engineers, and we don't really have any business experience outside of Wayne and Layne. We started with a few hundred dollars and a single kit, and we're seriously glad that crowdfunding was much rarer then! The mistakes we made just starting out would have been extremely expensive and damaging to our credibility, had our first order been for 10,000 units.

We seriously debated not going ahead with this deal, but after brainstorming a list of what we had to do to lower our risk, we felt OK going forward. Nothing was official at this point, so we pursued everything on our list that didn't cost money. There are only a few parts in the Blinky Grid and the Blinky POV, and they're basically common between the two kits. There are the LEDs—56 in the Grid, and 8 in the POV—but we aim to ship a few extra with each kit in case something is lost or damaged. Support costs for us to send an extra LED are much higher than simply including a few extra LEDs in each, especially when selling tens of thousands! There are the two light sensors, two resistors, a capacitor, a microcontroller, a socket for the microcontroller, a printed circuit board, and a battery box. The battery box holds two AA batteries, which we don't provide, and has a little power switch and two wires. Beyond that, the microcontroller needs to be preprogrammed with either the POV or Grid code.

**Figure 9-2.** *An enormous stack of parts means business is booming! Photo credit: Adam Wolf.*

Because we knew that schedule was extremely important, we made a big spreadsheet listing domestic and international suppliers of all our parts, and started sending out emails requesting prices and lead times on the quantities we'd need for 20,000 kits. As expected, almost all our international suppliers mentioned delays due to Chinese New Year, but our usual suppliers didn't even blink an eye at the number of resistors or capacitors needed. Our board house wasn't going to have any problems—their average order is closer to this sort of volume compared to what we normally ordered. There were plenty of light sensors, and a good number of the microcontroller in stock. We already had a good, trusted international LED supplier with plenty of capacity. The battery box was basically a generic, and we actually weren't able to find the original manufacturer. We got a variety of quotes from international suppliers, and quickly found out that for us, in our volumes, the cost driver of the battery boxes was going to be weight! They're relatively bulky, so getting them to the US was going to be expensive, compared to what the battery boxes cost. Our traditional domestic supplier for the boxes was much more competitive on cost than we had expected, but they had only a few thousand in stock. We had a thousand or so in stock ourselves, but we were still short by more than 10,000 units.

To figure out how long it would take to deliver our kits, we needed to know both how long it would take to get all the components in stock, as well as how long it'd take us to package them. To package a Blinky POV or Blinky Grid, there are a few steps. The microcontrollers have to be programmed, and put on a little piece of antistatic foam along with the IC socket. The populated foam piece and the capacitor, two resistors, and two light sensors go in the battery box. The battery box is placed in the retail tin, and either 60 or 9 LEDs are dumped in the tin. A printed card is laid on top of the parts, the box is closed, and then it needs to be shrink-wrapped. The retail packaging was something we hadn't dealt with before, but we didn't think it would be horrible. We took our time estimates from previous kitting sessions, added a good buffer for the retail packaging, and realized we needed to parallelize. With all the preparation steps (things like programming the microcontroller and cutting the resistor strip into pairs), and the packaging steps (things like shrink-wrapping the tins, putting those into labeled boxes, and putting those labeled boxes into other labeled boxes), we knew we'd be looking somewhere between one and five minutes per kit. Assuming we quickly invented a robot that didn't need maintenance or downtime, 1 to 5 minutes per kit for 10,000 kits works out to be 14 days of labor, all the way to 70 days at the long end! We also would probably have to use humans, who require breaks and food and time to sleep.

I live in Minneapolis, but at the time, Matthew lived in Pittsburgh. In order to avoid burning our money shipping parts and kits back and forth, we planned to ship everything to me, and Matthew would help from afar with the large amount of logistics and maintaining the rest of the business.

Like many kit businesses, we started out doing our own kitting, but sought help when orders increased. We hired my underemployed brother-in-law, paying him per kit as an independent contractor. After telling my parents about the potential deal, they instantly volunteered to pack kits. Matthew and I felt that there was still some risk there, so we looked around for more options. There were multiple fulfillment services in Minneapolis where we could drop-ship parts and have them kitted and sent back out, but the costs were completely prohibitive. I discussed our project with Jude, one of the people who almost inhabits our local hackerspace and helps keep it running. He seems to have worked with just about every technology and every person, and he pointed us to Opportunity Partners, a local nonprofit that provides opportunities to adults with disabilities—they do a lot of things, but their most revelant skill for us was retail packaging and kitting. I was a little skeptical—it would be very easy to run a shop like this and be completely skeezy, but after meeting with them and walking around on the production floor, I was completely sold. They looked like they had plenty of capacity and reasonable prices. Because we had never worked with them, however, we planned on splitting labor across all three sources: my brother-in-law, my parents, and Opportunity Partners.

Other things were simpler. When we had only a few hundred dollars of inventory, we didn't really worry about disasters. Absolute worst case—it kills our cash flow for a month. Increase that by 20,000 kits, however, and we'd be absolutely destroyed. A few emails to a local insurance company got us a quote for some business coverage with an extra rider for all that inventory for only a few hundred dollars a year.

It took a few days to get our enormous spreadsheet filled out, but we were able to add everything up, add some more contingencies in schedule and cost, and give our distributor estimated delivery times and costs. They asked for some clarification on some things, said everything looked good, and promised they'd get back to us shortly with an official purchase order after the retailer signed.

Over the next month, quantities shifted up and down, and eventually, the deal was officially canceled. Matthew and I did a small retrospective. We were glad we had had this opportunity to stretch our business and get some practice with large deals without losing anything. Then it was back on—with a caveat! We had to deliver faster, or else we'd miss the "spring reset." A "reset" in retail is when the layout of products in the store is rearranged. They were going to make some space for our two kits, but if we didn't have it ready in time, they'd have to delay or cancel the deal.

We looked at our schedules, all our quotes, our contingency, and gave the go-ahead. I mean, if we screwed it up, it was only our entire business and more money than we had dealt with in our entire lives that we were risking.

After getting the final purchase order, we started ordering all of our parts. The resistors, capacitors, and sockets were completely fine, and arrived in a few days. Our particular light sensor was no longer available in quantity at a reasonable price, but enough would be arriving at the distributors before our PCBs and LEDs would come so we bought what was available. After reviewing data sheets and verifying a sample, we ordered thousands and thousands of our generic battery boxes. Our calculations, which considered rush fees, showed that we could program the chips cheaper ourselves (rather than having the manufacturer do it), so we designed a few untethered programming rigs. All the kitter would have to do is place the chip in the zero insertion force (ZIF) socket, flip the arm, press a button, and it would be programmed. We quickly learned about shrink-wrapping, and ordered a few heat guns, sealers, and wrap.

**Figure 9-3.** *Adam's apartment began to get taken over by the boxes of parts. Photo credit: Adam Wolf.*

As the boxes started arriving, I realized I hadn't really thought about how much space this would take up. Wayne and Layne didn't have an office; we just ran it out of our homes, and my wife and I had actually already arranged to move to a new place in a few months because we didn't have enough space in our small apartment. Matthew and I custom ordered our LEDs to have shorter legs, because it's easier to make them fit in the tin with shorter legs and the legs are cut off once they're soldered anyway. Even with that, I had well over half a million 5 mm LEDs in my little apartment. Additionally, I had 415 kilograms of tins and 185 kilograms of battery boxes en route. Whoops!

There were a variety of little snags, like when our 85 kilograms of printed circuit boards were delayed in customs. After I reached a human at the appropriate department, I was told "The invoice says Foobar PCB Company, but the packing slip says FooBar Printed Circuit Board Company, so we can't release it." This sort of inanity isn't entirely unusual for customs. We were able to get it resolved with less than a week of total slippage.

As our parts arrived, we pipelined as much as we could. My brother-in-law and parents started prepping resistors by cutting the strips into pairs and programming chips. Not only did this help keep the schedule short, it meant I could unload some of the boxes from my apartment!

Eventually, we had enough parts for kitting to begin in earnest. The battery boxes still hadn't arrived, as they had a relatively long lead time, but we had more than 1,000 in stock. Not all our parts arrived in complete shipments either, but we had to make that spring reset—or else!

My parents got some parts, and my brother-in-law and my sister received the rest. They had a list of what to do, and some samples I had packed. We told them to notify us the moment there was an issue, and that'd we'd much rather be annoyed by them calling us about nothing than find out about something when we couldn't fix it in time. The rest of the parts, the microcontrollers, and the pallets of battery boxes arrived. I distributed them to both kitting groups. I met with Opportunity Partners again, and went over exactly what Wayne and Layne needed. We gave them parts for several thousand kits, and I went home nicely relieved that everything was going smoothly. Of course, only a few days went by before I got a late-night phone call from my father. Some of the tins weren't closing. After they had more than a few tins not close, he investigated, and found out that not all the battery boxes were the same. Some were a millimeter thicker than others, and some had a switch that stuck out farther than others. Why were we only seeing this now? I called my brother-in-law, and he said he hadn't seen it, but he hadn't really started packing with the new battery boxes yet. Aha! Checking the data sheet, the switch length wasn't specified, and the battery box thickness had a tolerance of a millimeter. Matthew and I hadn't realized the clearance was so close on the new retail packaging, and none of the samples had this issue. Oof. This was a serious problem —I knew offhand there were not enough battery boxes available domestically to replace all the new ones, and it would be nearly impossible to order internationally and still meet the spring reset deadline. Missing the spring reset, once again, would be bad news. Best case, it'd cost us a lot of goodwill—worst case, more money than we really had to lose.

After explaining this to my wife, we sat down on the carpeted floor of our living room in our pajamas and unpacked 7,000 battery boxes. At first, I thought about building a jig, but we were able to use our hands to feel each one for thickness and switch length, and sorted them into a "good" or "bad" pile. The things my wife does for love...did you know that two motivated people can inspect battery boxes for a one millimeter discrepancy at a rate of 3,000 per hour? A little over two hours later, we determined that with the pass/fail rates we were seeing in the boxes we had, we probably had to find only 1,000 battery boxes. While our main source didn't have that many, we found enough available domestically that we were able to burn some of our contingency budget and fix this problem. (We still have some of those battery boxes, ready to be used in a project where one millimeter of space isn't a deal breaker.)

After swapping the parts out at all the kitters, I double-checked to see how they were doing. Quality was fine, but the full retail packaging process was too slow. It was even slower than our worst-case estimate, with risk contingency on top! In order to get all the kits done, my parents had invited a few of their friends and my grandma to help. At this point, we proudly posted that some of our kits were "lovingly hand-packed by grandmothers in Wisconsin." Regardless of how awesome it was that we had grandmas packing kits, we needed to find the slowdown. After some analysis, it was pretty clear that we needed more scales for weighing LEDs and more heat guns; we easily solved this with a credit card and online retailers.

A week or two later, all the kits were packed. Some of the kits were already delivered, while others were in Duluth, Minnesota, and the rest were in Eau Claire, Wisconsin. You can fit approximately 3,000 boxed, tinned kits in a Subaru Outback (Figure 9-4). As they were being shipped freight, it was easiest to actually ship them out from the Opportunity Partners warehouse. The distributor arranged for pickup, a day or two later they were picked up and delivered, and we had successfully finished our first large retail shipment.

We had used all of our contingency, and a bit more than that, so our margins were a bit lower than they could have been, but even with that, it was an outstanding success. A few weeks after delivery, people started sending photos they'd taken of our kits in stores, saying they loved building our kit, had used it as a rainy day activity with their kids, and many other awesome things.

**Figure 9-4.** *You can fit 3,000 boxed Blinky kits in a Subaru Outback. Photo credit: Adam Wolf.*

My sister and brother-in-law used their check as the bulk of the down payment on their house. After paying our suppliers and kit packers, we had funding for years of Wayne and Layne experiments. Matthew and I developed a bunch of hardware for using Lego Mindstorms with Arduino, and wrote a book with John Baichtal about it. We wouldn't have been able to afford custom injection molding without the funding from this project. After that, Wayne and Layne took a few years to focus on services, instead of products. We partnered with a local company, and over the next two years produced dozens of museum exhibits and interactive installations that are located all over the United States. Today, we're looking for an office where we can have our very own laser cutter and pick-and-place machine, but more importantly, we've got prototypes of upcoming products out at the board house!

Adam Wolf is a cofounder and engineer at Wayne and Layne, LLC, where he designs kits and interactive exhibits. He also does embedded systems work at an engineering design services firm in Minneapolis, Minnesota. When he isn't making things blink or talk to each other, he's spending time with his wife and son. Photo credit: Adam Wolf.

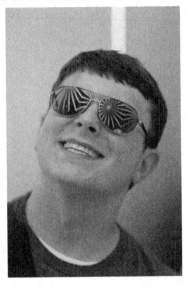

# Interview: MakerBot Industries Cofounder Zach Smith

WRITTEN BY **MIKE HORD**

**Mike Hord: So, how did you get started in this whole crazy Maker scene?**

Zach Smith: The real start, I would say, was in 2007. I had moved to NY, and was working as a web developer and started reading online about this new project called RepRap. I'm sure you're familiar with that; it was this group of people that said, "Hey, let's build a self-replicating 3D printer!" And all of those things were like, new, and kind of like new, mindblowing types of ideas. I started following it, I think I put their blog in my RSS reader, and just kind of watched it develop. I got more and more interested in it as I saw them continue to grow.

At that point in time, RepRap was just sort of a collection of wiki pages and support files; it was very early days of not really knowing the right ways of proto-typing and building and stuff like that. For example, when I started, the instruc-tions for making PCBs started at "download this file, and print it on a laser print-er, and iron it onto a PCB, and then soak it in acid." It was just "make your own PCB." And I did it, I tried it, and it was fun, but you know, I thought, there's got to be a way to do this.

I know that people manufacture these things professionally, and so I kind of made it my mission to figure out, all right, how do you make this, how do you make it better? How do you do it, you know, to the sort of, not professional standards, but you know, how do *real* engineers do this stuff? So I went out, and I looked around, and at that point in time there was this company Gold Phoenix, and so they had this, like, deal where you pay 99 bucks and they send you a crap ton of boards. But,

I only wanted *one* board, so I went on the RepRap forums and said, "Hey, I'm going to buy these boards. Who wants in on this, because I don't need 99 copies of this board?"

And from there, once I did that, I realized I kind of got sucked into making electronics because, as a web developer, I work in this digital medium where you design on the computer and then your work goes up to the server. It's kind of all digital. With electronics it was really cool because I could design on a computer, email it off to someone, and then a week later I get these very nice, professional things back. So, that really appealed to me.

Really, what got me into 3D printing and digital manufacturing kind of stuff was that it kind of reduces the physical objects to this software-based design where really the true essence of what you're building, what you're making, is digital. You're using your computer and then a robot is making it for you. You know, I consider myself a Maker, but I don't have very high skills when it comes to building things by hand. I can put things together, I can bolt things, use a screwdriver, various things like that, but with the kind of tools you have available now between 3D printing and CNC and working with PCB fabs, the level of precision and quality that you can get is amazing, it's just kind of...mind-blowing. How high quality can you get? So it really kind of frees your creative side from your ability to implement that creativity. So, that's one of the things that really drew me to 3D printing and digital fabrication in general.

**Mike: So, this order that you did with Gold Phoenix; this was in 2007? 2008?**

Zach: I'm not sure exactly when it was; it was somewhere in that time range. I would say, maybe 2007.

**Mike: And at that time, your full job was...well, to get a sense of background, how old were you then? How old are you now?**

Zach: I am 30 years old now; at that point in time, seven years ago, I would have been 23-ish.

**Mike: And you were working as a web designer at that time?**

Zach: Yeah, I worked for Vimeo back during that time.

**Mike: OK, so your background at that point was entirely in programming?**

Zach: Yeah, all the mechanical and electronic stuff was learned on the fly. Reading through all the Arduino tutorials, all the different websites I could find. During those very early days, I had a small apartment in Brooklyn, kind of a college dorm style apartment. I had a lofted bed, and so I would sleep in the top part of it and in the lower part I had a tiny little desk and just a soldering iron and stuff like that. It was fun, but it was not a very ideal sort of situation. I started reaching out to the different people in New York that I would see in *Make:* magazine. In particular, there were Phil Torrone, LadyAda, and Bre.

Sometime in 2008, there started to be what we called the microcontroller study group. It was just kind of a weekly meeting where we'd meet at a coffee shop, and later someone donated a space for us. People would bring their projects and we would ask questions; we would just kind of geek out about electronics and robots and whatever people were interested in. So, through there, that kind of evolved into NYC Resistor, which was such a huge step and such an important part of this journey from curiosity about this thing to making it into a sustainable business and something that I could do full time, because it gave me a place to work.

New York is very expensive, and we were able to combine resources. Everyone had sort of their own interests, but we all had this shared interest in making things, we needed a place to have a drill press, and later, a laser cutter, and more things like that. Things that I would not have been able to afford on my own. We were able to put that together, and so I had my own little corner where I would have the RepRap parts that I would ship out. It also was a good place to meet people. There was a nice culture of making and just a really nice environment for that kind of stuff. I guess the next step in the whole saga, in 2009, we went to the Chaos Communication Congress. A bunch of us from NYC Resistor went there, and it was there that Bre, Adam, and I decided to start MakerBot.

That was a really exciting time for me, and for them as well, and we decided, "Hey, you know, let's see if we can make a business out of this!" Before that, I was sort of selling parts online, but it was more out of kind of a need for myself to get these parts, because so many of these things you build, there are economies of scale that begin pretty soon but it doesn't make sense for an individual to buy that many things. For example, I went out and found a plastics supplier, and they were like, "All right, sure, we can do this for you, but our minimum order is 50 pounds of plastic." At this time, I had an extruder that barely worked, so the idea of 50 pounds of plastic was just crazy.

**Mike: That's your lifetime's worth of plastic for that device!**

Zach: Yup. So I go on the forums, and say, "Hey, I found the place that has plastic. Does anybody want some?" The same thing with the PCBs, motors, belts, and pulleys. The random parts that you can't get at a hardware store, that you can't buy single parts of. It kind of grew from there and we realized there was a need. We were all very passionate about trying to make this more accessible, make it easier. When we started, our long-term goal was, "Let's make a really cheap 3D printer that you can just buy, open, and use." Anybody should be able to use it.

I'm pretty proud that MakerBot has been able to achieve that; I can't say I necessarily agree with the route that it's taken there, but the company did achieve that goal that I initially set out to do. When we started MakerBot, originally we had some very lofty goals: we wanted to be open source, we wanted to be friendly with the community, we wanted to build awesome 3D printers, we wanted them to be as cheap as possible. It was, for my time at MakerBot, it was a really good time. There's all sorts of drama that happened. I don't know if you want to go into that; I've told that story so many times it's kind of like picking at scabs now.

**Mike: I'm not going to put you through that. I'm interested in the arc that carried you into that space, and then the reasons you moved past that. For now I'm curious about what your experiences were like as you moved into that arena. Did you find yourself rapidly feeling in over your head? Did you feel that it was far more business, and shmoozing, and executivism than making things? What was that like for you?**

Zach: I don't think there was a lot of...the business side wasn't really a problem. Even when were up to 100+ people, the business stuff...we sold online, we would advertise it at events. It wasn't like we were cutting massive deals with companies where the corporate stuff comes into play. For me, the feeling of being in over your head was, all right, I want to build something really awesome, something I'm really proud of. I was a web developer and I taught myself how to build electronics and hack together a prototype. How do I "level up" and become a professional at this? How do I use the right techniques, the right designs?

For me it was kind of this scramble to learn these things that you would've been taught at a university getting an engineering degree. So, for me it was this scramble to get an appropriate level of professionalism and build that into the product. Part of that we did by hiring engineers, but when you're a small company, it's kind of hard to attract that A-level talent. We did hire a bunch of really good engineers, but a lot of the people were entry-level type people who maybe had a few years under their belts, maybe had gone through a project as a junior engineer, but nobody could come in and was like, "OK, here's how you take a product from idea to final production and shipping and make something that's really amazing."

So, a lot of it was learning by trial and error, building things, listening to customers, dredging through emails that are like, "Hey, this thing broke after 20 hours! What the heck?" Learning the difference between making one thing that works for you in the lab and making 1,000 things that are going to work 99.9% of the time for other people. That was my big struggle: here's this cool idea, but is it going to work for everyone? Is it going to be something we can reliably produce multiple copies of? I think that's something that a lot of companies really struggle with.

Post-MakerBot, I work with a hardware accelerator in China here called "HAXLR8R." Of the companies that make it through the prototyping stage, they spend at least the same amount of time, usually longer, working with their factory to iron out the bugs in actually producing their products. It's just such a monumental type of thing because so many different things can go wrong during the assembly and during the manufacturing of your products. Any potential point of failure, you're *going* to hit that. Someone's going to put a wire in backwards, or someone's going to not tighten something down all the way. There's all sorts of things that could go wrong. Learning about crap like tolerance stackups and designing with that is...you know, in the lab, if something does fit, well, grab a file and file it down, and cool, it works! But that just doesn't work if you're trying to make a large number of something.

**Mike: I hear you there. That is the universal struggle. Making one thing on the bench is easy, making something that works for 10,000 people is a much harder challenge. And that is a painful thing to learn, for a Maker, so I'm glad you mentioned that. So let's move forward a little bit here: you've finished your time at MakerBot, what did you do after that? I don't want to get into the nitty-gritty, the hairy details of it, but at some point, there was a point where there was going to be a parting of the ways between you and the other founders. So tell me a little bit about that experience, that moment of sitting down and saying, "Oh, great, what am I going to do next?" There had to be some level of a feeling of freedom, but also some fear. What was that like for you?**

Zach: I was basically fired from the company that I started. There were a lot of things that went through my head; for about a year, and maybe even two years, I definitely went through a little bit of depression. It was tough. I had very tightly defined myself as "I started MakerBot; I build 3D printers," and so I went through an extended period of soul-searching, and what do I want to do, and who am I? I tried a bunch of different things, and I had a very negative outlook on the future of MakerBot. I didn't think that my partner, Bre, had the skills to build this company up. I guess I was pleasantly surprised when they sold the company; I really wished they had stayed independent. Selling to Stratasys and going closed source are kind of both things that are the antithesis of my vision for the company, but that's not really something I could change at all.

Part of the whole blow-up situation was we had come to China and just kind of disagreed about how this project was going. There were issues; when the

MakerBot team decided to shut down the China project, I didn't take that very well and that kind of precipitated the whole fallout mess. But, I had been out in China maybe nine months and found it to be this incredible place. It's such an adventure out here. Especially if you're a Maker! A lot of people call it the factory of the world, and it is. There are so many factories out here, and because of that there is this massive ecosystem around it of suppliers and markets and places where you can go get stuff, and it's all really cheap and everything happens really fast. So I wanted to stay out here and learn more about that.

Also, I have this feeling of incompleteness, in that we had started this project, we were really close to manufacturing, then it just got the axe. I felt like I hadn't finished what I came out here to do. I wanted to come out here and continue learning and continue to grow and understand more about how this part of the world works and how manufacturing works and so I decided to stick around out here. It's been awesome; I've found China to be a very welcoming country. It's really awesome if you like making things: in my neighborhood, I don't really need a laser cutter, because I can hop on my electric bike and go a couple of blocks and there's a dude with a laser cutter. I live in the middle of downtown, and there's parks, and it's nice. A lot of people have this idea of China as this toxic wasteland. Granted, maybe it's like that in certain areas, particularly if you go up into the north, but Shenzhen for whatever reason of geography or climate has very little pollution. So you have this combination of high-tech manufacturing, markets, cheap supplies, fast prototyping and general willingness to do stuff. I just decided to stick around here and it's been a blast so far.

The main thing I've been doing out here is working with HAXLR8R. We've gone through about four batches of companies. It works very similar to Techstars. There's a lot of incubator, accelerator-type things out there. Basically, what we do is bring 10 companies into Shenzhen. We have an office that's full of all sorts of prototyping gear. It's located in one of the giant electronic markets here in Shenzhen. We let people build their ideas for three and a half months, and along the way we provide whatever help we can. We'll have like a sit-down with mentors: what's your business plan, what's your marketing plan? What's your vision for the product? What's your industrial design? We'll bring people in to help them with all those different things.

For me, it's fun because I love this stuff, and it's awesome to get in there and kind of work on a bunch of different projects. It's also great because I still feel like I don't know a lot about this stuff. I mean, I guess I've been doing it for a long time now, but it's like the old adage: the more you learn, the more you know you don't know. For me, Shenzhen is the perfect place for that because there is so much here that is different. It's so easy to learn something new every day, because between the normal stuff like manufacturing and technology, you can walk through a market and see new things. If you want to go check out a new manufacturing technique, you just look up a factory and say, "Hey, I want to come tour your factory!"

One of the techniques I wanted to see was thermoforming, taking a sheet of plastic and heating it up and stretching it over a mold. So I looked up a couple of them on Alibaba, scheduled a visit, went and saw them. I was able to learn a bunch about what the potentials are. I used to have this book in New York called, oh, I can't remember what it's called. It's a really awesome book, it's a coffee table book, and it just shows you hundreds of manufacturing techniques, and I would just flip through that and sort of drool over all these things. When I got out to Shenzhen, I realized, "These things are all right here, and the factories are all really accessible and willing to give you a tour!" So I kind of just went through the book and said, "Oh! I want to see that one!" and then I went to go see. It's so different to see it in real life, how it's actually being used, instead of this sanitized block diagram in a book. To be able to ask the guy as you walk through "What's that? How do you make that? Why do you have plaster ones, and aluminum ones? What's the price difference, and the quality difference?" It's just, it's a really interesting way of doing things out here. It's just really accessible.

**Mike: So, what I'm hearing is that, to the Chinese, the concept of open source is just...how it is. It sounds like they're really open about what they're doing.**

Zach: Well, a lot of these factories don't produce original designs. They're there because they have skills with a particular technique, right? They're very good at taking this particular manufacturing technique and doing it. They have the machines, they have the knowledge, they have the workflow, they have the processes. So, for them, taking a client through their factory is promotional material. They're saying, "Hey, here's what we do. You can see that we're skilled at it, you can see the actual results of what we do!" And they're very open about that because hearing from a guy for a couple of hours about how this all works and what they're doing is so much different from opening your own and implementing that.

So, in that sense, they're very open about how their process and stuff like that works. There certainly is an open source movement in China; there's the whole shanzai, which is the copycat culture. There's also what bunnie Huang has termed *guang-kai*, which is kind of open source, but it's from—and I'm not an expert researcher into this; I think it's interesting but I don't have the language skills to dig into it—as far as I know, it's more like reference designs that a certain company will put out and then other people will manufacture and it's because those companies want to sell those chips, they want to sell the modules, they want to sell these different things. So they'll put it out there, and you have this open source-ish thing, but what I think it's missing is "we're doing this out of an altruistic type of motivation," but it still has this open source thing where you can get in there and see how stuff works. I think that's pretty interesting.

**Mike: So, it kind of lacks that "improve me and pass along the improvements," "shoulders of giants" thing you see in the open source movement.**

Zach: Well, you do have open source companies though. Seeedstudio is a great example. There's a few others; there's a robot arm company here in Shenzhen called uFactory; I'm actually going tomorrow to meet with these guys who went through HAXLR8R. They're called "FlexBot" and they make these little quadcopters that are yay-big [indicates palm of hand] and they primarily use digital fabrications, like CNC and 3D printing. They've apparently set up an automated line where they have a bunch of 3D printers running and they've got a robot arm that unloads it and then they'll start up a new job; I'm going to go check that out. I think more and more people are starting to understand the other side of that, and the intangible benefits and the ethos behind the Open Source movement.

**Mike: How long have you been in China now?**

Zach: About three years, maybe two and a half if you go by total logged time.

**Mike: Is there anything else you'd like to share? Anything you want to make sure gets on the record?**

Zach: I'm not sure. It's hard to explain the magnitude of what it's like to be building stuff and out here. The scale is just so much bigger. I mean, I was back in Austin for a few months, here, and I went up to the TechShop and learned to use a bunch of the machines, which was really cool, but then trying to find parts and stuff, I was back to ordering everything online, which is kind of a pain in the butt when back here I can just go down to the market and go shopping and get everything I need.

There's a few places here, there's one market here that is my new favorite one. It's literally everything you need to build a factory in one building. On the first floor they have normal electronics: everything from soldering stations to all the tools you need to build stuff, like tweezers, magnifying glasses, soldering irons, chemicals. You have components, people with prototyping PCBs, everything. Then you have the heavy equipment, lathes, drill presses, welding equipment, inspection machines, micrometers, and calipers. All kinds of crazy stuff. They were selling these automated wire-cutting and stripping machines for 500 bucks. And then you go up to the second floor and it's motors, and stepper drivers, and people selling all sorts of pneumatic equipment. There's this one booth selling all the different heads and tips for pick-and-place machines. People selling all sorts of ball threads. I was going through and thinking, "Oh my God, I could build a 3D printer or CNC machine with just the stuff I find in this building!"

I don't know. One of the things I really like about that sort of accessibility is that as someone who didn't go to school for any of this stuff, the important questions to answer are: *What is possible? What is accessible? What do you not need to design?*

I've seen a lot of projects and a lot of things where people reinvent the wheel a lot. Going into one of these places, you can walk around and in half a day, get 20 ideas and realize, "I don't have to reinvent this wheel! I don't have to design something from scratch because I can get these module pieces off the shelf and put them together!" You can get that from an online catalog, but there's something different from being able to see it in person and pick it up and look at it and ask a person the question of "Hey, what is this? How do people use it?" That's the sort of thing you don't really get out of a data sheet. You get load factors and operating parameters, that, if you know how it's supposed to be used, are really super helpful, but if you're asking the third-grade question of "What the heck is this widget and what am I supposed to do with it?" it really helps to be able to see it in person or ask someone. I think that's one of the less frequently discussed advantages of having things that you can go see in real life. I would highly recommend that Makers come over here to check it out. If nothing else, just to broaden your horizons and see how a different culture approaches the same sort of idea, the topic of making.

**Mike: That's very cool. One last thing I'll ask: Where do you see yourself in five years? Any guess, anything about where you'd like to be? How do you think things are going to go for you?**

Zach: I'm not sure. I don't have that...when I was 25, and building MakerBot, I kind of had that laser-focus of "this is exactly what I want to do." I like think that I'm back into like a student mode; I'm learning things. I'm studying Mandarin, I'm really interested in robotics and robot arms and control theory. I'm just trying to stay open to what that next obsession is going to be. I do know that I want to stick around in China for a while. My plan for later in this year is, if you go on the outskirts of town, you can get extremely cheap factory space, like $1.50 a square meter, which is like 9 square feet. So I'm just going to get something that's like five to ten thousand square feet of industrial space with giant freight elevators, and

I'm just going to get a bunch of tools and build stuff. It'll be fun! I'm going to start looking for that space in the fall, when my lease is up on my current workshop.

**Mike: I'm not going to lie to you: I'm a bit envious of your freedom to do that! I expect we'll be hearing some pretty fun things from you out of there.**

Zach: I hope so!

---

Zach Hoeken likes to dream big, fail big, and win big. His true passion in life is acting as a catalyst and helping others do amazing things.

Whether it is creating open source microcontrollers, robot controller software, object sharing websites, or 3D printers, there is one central purpose: to help other people help themselves create an awesome world to live in. That's why he built MakerBot.

He hopes that someday we can create a world that surpasses even the wildest futures portrayed in science fiction.

He thinks the universe is and will continue to be completely rad. Photo credit: Zach Smith.

Magnificently intelligent and dazzlingly attractive, Mike Hord is also an accomplished liar. After five years of corporate drudgery, he shook free of his bonds and began work in the SparkFun Electronics engineering department, where he designs boards, writes tutorials, and interrupts during meetings. An active member and passionate proponent of the open source hardware community, he spends his free time teaching his two minimakers how to void warranties and indulge curiosity. Currently there is no expectation of release from SparkFun, as he has not yet shown any signs of good behavior. Photo credit: Pat Arneson.

# Make a Living Doing What You Love

WRITTEN BY **MITCH ALTMAN**

I make a living doing what I love. I highly recommend it. It's quite wonderful. Well worth a try.

It wasn't easy for me, but I'm really glad I put in the time and effort, because being depressed sucks. I know. I spent the first half of my life there. As an attempt to numb my pain, I sat in front of a TV during most of my waking hours. I was a total geek, so I also spent lots of time working on solo geeky projects and learning electronics. These activities further fueled being bullied at school, made me more depressed, and I withdrew into TV and geeky projects even more when I got home. I blamed myself for all this. Sigh. Feel free to ask me about all this sometime, since I'm happy to share. But this essay is about how cool it is to live a life doing what you love—so, let's skip ahead a bit.

By 1993, I had managed to transform my life from being a total depressed blob of a kid to being OK with who I was. I had quit TV cold-turkey many years before, but kept up with being a geek, making a decent living as a consultant by helping small companies with their microcontroller projects. I got paid to do a bunch of projects, such as helping make museum exhibits, virtual reality systems, disk drives, and voice-recognition systems—all things that I thought were pretty cool.

By 2003, 10 years later, my work was still pretty cool, but it started to wear on me. I didn't want to put all of my energy into stuff that was merely pretty cool—I wanted to put my energy into what I truly loved! What would life be like if I could do that? I didn't know, I wanted to find out! I made time to explore by saving up a year's worth of expenses. And I decided that for one year, I would choose to do something only if I loved it. If I loved the idea of helping a friend move apart-

93

ments, I'd do it. If I didn't love the idea of going to a *Star Trek* movie marathon with a bunch of friends, I'd use the time to do something else. If someone called wanting to hire me to do a project that was pretty cool, but I didn't love, I'd pass. That was scary. How would I make a living? I didn't know, but I knew there had to be a way. There must be a way to do what I love, and by doing what I love, make enough to keep doing what I love. I couldn't think of a better definition of success.

**Figure 11-1.** *Mitch Altman teaching a workshop at the OHM2013 conference in Heerhugo-waard, the Netherlands. Photo credit: Mitch Altman.*

I didn't work on TV-B-Gone to make a living. I just wanted one. Of course, it turned out that all of my friends also wanted one. And my friends told their friends, many of whom wanted one. And when it turned out that many of my friends' friends' friends also wanted one, I thought, "Hmm, maybe there's an opportunity here."

To start my explorations, I used my time to do a bunch of volunteer work that I totally loved. I also started working on projects that I'd been thinking about for a long time. For years and years, after working with electronics all day, I wasn't so motivated to play with electronics afterwards. But now I had time and energy, so I started working on a project that I thought up years earlier: in 1993, I was trying to pay attention to my friends at a restaurant, but was distracted by a TV blaring away in the corner by the ceiling. I noticed my friends staring at the thing, too. We didn't get together to watch TV. We came together to talk, catch up, and have a nice meal together. Yet, we were all unable to keep our eyes away from the TV. I asked why there wasn't a way to turn these annoying things off, and I instantly realized I knew a way! And I blurted it out: I was going to make a device that would send out OFF codes for every popular TV made, one after another. One of my friends gave it a name: TV-B-Gone—a one-button universal remote control that would turn off TVs in public places. And years went by...

But once I started working on it, TV-B-Gone (Figure 11-1) took on a life of its own. I became obsessed with finding all of the remote control OFF codes for every TV out there. It took a lot more work than I expected, but I didn't care. I wanted to turn those TVs off! After a year and a half, I finally had a working prototype. After a lifetime of being controlled by TV, now I had a way to have power over all of them. And I went all over San Francisco, enjoying the hell out of turning TVs off everywhere I went.

I had no idea then, that this project would soon change my life forever.

I didn't work on TV-B-Gone to make a living. I just wanted one. Of course, it turns out that all of my friends also wanted one. And my friends told their friends, many of whom wanted one. And when it turned out that many of my friends' friends' friends also wanted one, I thought, "Hmm, maybe there's an opportunity here." I decided to take a gamble and make as many as I could afford, which was 20,000. I calculated that I could break even if I could sell 5,000. Even if it took several years to sell them, I'd be fine with that. There would then be 5,000 people going around enjoying TVs off all over the world! And any more would be gravy. But, as it happened, I sold all 20,000 in 3 weeks! And it is the only way I've made money since 2004. It seems that this is the kind of thing that can happen when you do what you love, simply because you love it: if you love something, chances are others will too. And in capitalism, if people love what you do, they will pay you to do it! Worked for me. It works for lots of people. Maybe it could work for you too?

Before the first day of TV-B-Gone sales, I stayed up all night with my friend Chris, getting my website together. We wanted it to accept credit cards, just in case a few people might want to buy a TV-B-Gone remote control after an article about it on the WIRED website (http://wired.com) was published. That article was the result of an acquaintance, with whom I volunteered, interviewing me about TV-B-Gone as a final project for his journalism class. We had a fun time going around San

Francisco turning TVs off, and he'd asked people what they thought about it. He wrote it up, pitched it to a few places, and Wired.com picked it up. Chris and I made the website live at 5:05 a.m., and thought there must be a bug in our shopping cart, because we instantly received emails alerting us to sales. But, no, they weren't bugs, orders were already coming in! I had an overnight hit.

**Figure 11-2.** *Students learn to solder from Mitch Altman in a class in Shenzhen, China. Photo credit: Mitch Altman.*

At 9 a.m., National Public Radio called to interview me. I guess I did OK with that interview, despite the lack of sleep: after it aired that morning, the phone didn't stop ringing. *New York Times, Reader's Digest*, Deutsche Welle, BBC, Radio France, CBS, NBC, ABC, CNN, and on and on. I agreed to be interviewed by all of them. And, as it turns out, I love being interviewed. Journalists ask me questions, I can state my thoughts and opinions, and they're sent out into the world. What's not to like? Again, this happened simply by doing what I love.

If anyone had told me that TV-B-Gone would lead to me giving public speeches, I would have done everything in my power to destroy the project. Like most people, public speaking terrified me. (It still does.) But, because of the media interviews, I was invited to the first Maker Faire, and to my first hacker conference (HOPE 6, in New York City) to give a talk. At these events there were thousands of introverted geeks of all sorts. I felt safe and wonderful being in a group for the first time in my life. I found my tribe. It was wonderful! And, it turns out that I loved public speaking. I'd stand in front of a crowd of people, I could (and would) state my thoughts and opinions, and everyone would sit and listen, and laugh at all the right places. What's not to like? I've been doing it ever since. These kinds of things seem to happen when you do what you love.

**Figure 11-3.** *The TV-B-Gone turns off nearly every TV on the planet. Photo credit: Mitch Altman.*

I noticed something odd, however. At my first my first hacker conference, and even at the first Maker Faire, no one was actually making anything. Since they were so cool, I knew I wanted to be more involved at these events, and I thought of a way to do it. I wanted to make some cool, easy electronic kits for beginners, set up a table with a few soldering irons, and teach soldering at the next Maker Faire and at my next hacker conference. I did. I loved it, and I was mobbed with people loving to learn to solder! So, at my next event, I made a lot more kits, and set up more soldering irons. And it was way fun! And I was mobbed. At the next event, made even more kits, and set up even more soldering irons. And I loved it. And I was mobbed with people wanting to learn to solder. This soon grew to a huge area of 50 soldering irons, with an army of enthusiastic volunteers whom I trained to teach soldering, and we could teach over 3,000 people of all ages to solder in a weekend. And no matter how many kits I made, they would sell out. Clearly the world needs more of this type of thing. Again, this was an outgrowth of doing what I love.

I gave more workshops! But it was always sad when a Maker Faire or hacker conference came to an end (Figures 11-1, 11-2, and 11-3). Then I'd have to wait until the next one. But, at my third hacker conference (Chaos Camp 2007, in Germany), three German hackers gave a great talk on how to start your own hackerspace. Hackerspaces, as they described them, are physical places, with a supportive community for people to do the sorts of things that geeks do, the things people do at hacker conferences and Maker Faires: computers, tech, software, all sorts of arts and crafts, science, food—anything that people at the space love to explore and do, teaching, learning, and sharing. I was inspired to help start a hackerspace in San Francisco, where I live, so that I could have much of what Maker Faires and hackerspaces have to offer, but in my hometown, and all day, all night, all year round. Many others were also inspired to start hackerspaces in their hometowns. Some friends and I started Noisebridge hackerspace, and others started hackerspaces where they lived. And we helped each other make them happen. It was a lot of work, but since I loved it, it was way worth it.

I didn't set out in life to help start a hackerspace movement. But that's what happened. Once hackerspaces started in a few cities, they were so cool, and helped so many people, that people started them all over the US and the world. And whenever I was invited to give a talk or workshop at a conference, I looked up the local hackerspace, and offered to give workshops there. They were always very popular and lots of fun. And if I went somewhere without a local hackerspace, I'd encourage people I met to start one. Many did. People need community, and people need to express themselves creatively. Hackerspaces provide for these two deep human needs. And hackerspaces have been growing exponentially since 2007. There are now over 1,600 listed on *http://hackerspaces.org*, a central networking website where hackerspaces all help each other. This was a result of doing what I love.

As it turns out, the more workshops I give, the more I'm invited to give. And, because I love giving workshops at hackerspaces around the world, I go around the world giving workshops at hackerspaces! I don't make any money giving workshops, but I don't lose any, either. The result: I get to travel around the world for free, meeting the most amazing geeks everywhere I go. And we all help each other. Some fabulous additional benefits of living a life I love.

**Figure 11-4.** *A soldering workshop at Netz39 hackerspace in Magdeburg, Germany. Photo credit: Mitch Altman.*

It also turns out that by traveling the world and giving workshops and talks, I became well-known in lots of places. This is advantageous for hackerspaces: when I give a workshop, many people who show up have never been to a hackerspace before, and find that they really love it! Hackerspaces grow as a result. And lots more people learn to solder, make cool things with electronics, and gain confidence from their experiences. Many people also choose to make TV-B-Gone remote control kits and then enjoy turning TVs off, which makes the world a better place everywhere they go—always a bonus. These things really do happen when you do what you love.

As this movement has grown, the world has changed a lot. Libraries have started hackerspaces to attract more people, and provide even more valuable services for their communities. Museums give more workshops and allow people come and use their tools. Schools and universities have started hackerspaces to provide students opportunities for hands-on, play-based learning, which is an incredibly effective educational method. I get to give workshops at all of these places—anywhere and everywhere that people want to learn. And I love teaching.

At this point in my life, I continue to do what I love, and make a living from it. I make enough doing what I love, to keep doing what I love! This is still my definition of success.

I don't know what's next for me. No one can predict the future. Things keep changing. I keep changing as a result of the choices I make. All I know is that I am going to keep making choices based on what I love.

Please know that it is an option to make a living doing what you love. It really is. There are other options, of course. You can, if you like, make a living doing what you don't like. A lot of people seem to choose this option. I'm not really sure why. You can even, if you like, make a living doing what you hate. Many people actually choose this option. Nothing wrong with it. But perhaps it's worth exploring other options? It might be kind of scary. Things might turn out very different than you expect. But perhaps it is worth a try? It's totally up to you.

---

Mitch Altman is a San Francisco–based hacker and inventor, best known for inventing TV-B-Gone remote controls, a keychain that turns off TVs in public places. He also cofounded 3ware (a Silicon Valley RAID controller company), did pioneering work in virtual reality at VPL Research in the mid-'80s, and created the Brain Machine, one of *Make:* magazine's more popular DIY projects. He has contributed to *Make:* magazine, written for *2600: The Hacker Quarterly*, and for the last several years, has led workshops around the world, teaching people to solder, make cool things with microcontrollers, and promoting hackerspaces and open source hardware. He is also cofounder of Noisebridge, and President and CEO of Cornfield Electronics. He is about to launch his latest project: NeuroDreamer sleep mask, to help people sleep and have lucid dreams. Photo credit: Mitch Altman.

---

# 12

# Are You BioCurious?

WRITTEN BY **ERI GENTRY AND TITO JANKOWSKI**

Biotech isn't limited to labs anymore. There are cool hardware and low-cost hackerspaces springing up and being used by beginners and scientists. With this influx comes new ideas and new applications.

## DIYbio

DIYbio, Do-It-Yourself biology, takes some unpacking. Telling people about "hacking biology" evokes many emotions: disbelief, amazement, and questions about personal safety and national security! Some wonder what sorts of miscreants would practice science in their homes (aka their dark, dingy basements) and some want to get their kids involved ASAP.

What a strange new world, this DIYbio! Literally, it is part DIY and part biology (also: hardware, software, community, and policy—but we'll get to that later). The people involved in this space sometimes self-identify as scientists, but might be better known in their civilian lives as artists, engineers, high school students, high school teachers, Makers—you name it!

When most of us think of scientists, we reasonably draft images of white-coated ladies and gents in glasses working in a sterile environment: the lab. This lab is likely in a university or industry setting, where serious scientists methodically design experiments that could change the world, or at least help us understand it.

The old concept of scientist works, but is a misconception. It's what happens when science isn't a part of our everyday lives. We can become so far removed that we lose touch with our inner scientist. We've heard it before with art: ask any classroom of six-year-olds who likes art, who can draw, who's an artist. Everyone raises their hands! Try that with a room full of post-educational-system adults, and how many dare to put their hands up? The same is true with science. Take a moment and think back to an early memory of science. What was cool and exciting about it?

DIYbio is all about making those awesome parts of scientific discovery available to people. Call it open science, democratization of science, DIYbio...whatever you will, it starts with a mindset: believing that science isn't constrained to institutions. And, like many movements of today, DIYbio started with a community.

DIYbio (*http://DIYbio.org*) was founded by Mackenzie Cowell and Jason Bobe in 2008, first as a set of meetings around MIT and Harvard, and then as an online network of people from all over the world. In its early days, nearly half of the listserv was artists. When you have such a diverse group of people trying to communicate their ideas around science, you can either achieve chaos or watch as a sort of common language of science starts to evolve.

A big success factor of DIYbio is that it's allowed anyone—from any sort of background, scientific or not—to get involved in the conversation. Someone with only a partially formed idea, say, for building an open source gene sequencer, can jump on the forum, ask for ideas, and (if it's sticky) a sort of collaborative discovery begins.

## OpenPCR

"Is that an Arduino?" squeaks a blond fuzzball. It's a little kid, wandering around Maker Faire with his dad. He peers inside the OpenPCR DNA Copy Machine (Figure 12-1) and points at that blue board inside. "Yeah, it is. We're using the Arduino to play with DNA!"

You've used an Arduino to make a blinking light, right? You sat at your computer, typed some stuff, and practiced not burning yourself with a soldering iron. We did the same thing to build OpenPCR. We sat at our computers. We typed a lot of code. We soldered a bunch of stuff together. Sure, it was more complicated, but fundamentally it's the same type of work.

Hardware, physical "stuff," is a great way to get interested in biotech. DNA itself is invisible to the naked eye. Billions of DNA molecules fit in a drop of water the size of a lady bug. It's hard to get excited about a drop of water. That's where hardware tools come in. Hardware tools like OpenPCR make biotech tangible. You can hold an OpenPCR machine under one arm. You can put it in your backpack. If you come up to me at a Maker Faire, and I'm holding an OpenPCR, I'll hand it to you. Now you're holding a DNA copy machine. That's kind of cool! "Something something-something copies DNA something-something," I'll explain. And it's in your hands! It's kind of like holding a box of kittens. You're standing there with it, and you kind of want to set it down, and you're not really sure how to safely set it down without breaking anything.

**Figure 12-1.** *The OpenPCR project brings a high-end tool to amateur hands. Photo credit: Tito Jankowski.*

Want the secret to learning to draw beautiful sketches? Take a coffee cup from the kitchen and put it in front of you...upside down. The upside-down part is the key, because something magic happens when it's turned upside down. When it's upside down, your brain doesn't see it as a coffee cup anymore. The cup is just a bunch of blobs, shadows, and colors. Your hand takes over. On the paper appears a beautiful drawing of a cup. If you leave it right-side up, your brain gets in the way. You end up thinking about all the coffee cups you've seen before (big mugs at Starbucks, dirty cups at diners, and that talking cup in *Beauty and the Beast*). Your lovely brain leaps at the chance to capture the spirit and soul of the cup. To be or not to be, what *is* a coffee cup? All that thinking gets you a bland coffee cup drawing. But when it's upside down...beauty flows (Figure 12-2).

**Figure 12-2.** *Sometimes magic happens when something's turned upside down. Art credit: Theodore Jankowski.*

OpenPCR is like that upside-down coffee cup. By turning biotechnology up-side down, you can get a better grasp on it. Making it an open source kit that you can build at home is like turning the cup upside down. There's lots of devices trapped in labs right now, with a label on them that screams "don't touch me!" OpenPCR looks more like a toaster or a bird feeder than a fragile, expensive lab instrument. Cocreators Tito Jankowski and Josh Perfetto designed it from scratch with $12,121 raised on Kickstarter. Maybe your mom has one, or your friend has one in her school locker, or maybe there's a scientist using one on a submarine coasting around the bottom of the ocean. By bringing in beginners and new minds, we can really flip our understanding of biotech and refresh our preconceived notions. DIYbio, OpenPCR, BioCurious, and all the other wonderful things going on are about turning over the coffee cup. Forget anything you know about biology, science, and biotech. Come look at biotech with the eyes of a beginner.

## BioCurious

BioCurious is a working laboratory, technical library, and meeting space where entrepreneurs and the merely curious can learn about biotechnology.

Think of it as the Minecraft philosophy of life.

If you haven't played Minecraft, it's kind of like playing with Legos on your computer. There are a few basic blocks, and with these blocks you can build anything. There are only a handful of basic blocks, such as water, dirt, and grass. But by combining these blocks, miners have built castles, slides, and even a working computer processor within the world of Minecraft. I think life is much like Minecraft. As people, we're all made of a few emotional ingredients: passion, love, fears, and doubts. And we all become different people—artists, teachers, scientists, family doctors, Uber drivers, and tax accountants. Everyone is built from a few "blocks" of emotions, just like the world of Minecraft is built. What's cool about the world of biotech is that it's kind of a new "block." That is, you can integrate biotech into whatever you're working on.

Are you a car fanatic? Drive to BioCurious and let's talk about how to tweak your engine to use biofuels. Or dream together about a living paint that molts so that every month you get a brand new shiny paint job on your car. Or maybe you're a foodie. Want a slow-cooked filet mignon and perfect poached eggs? Let's be really meticulous about the temperature and cook a delicious meal with Nomiku's new sous-vide cooker. We all know how thrilling a full belly feels. Napa Valley wine and cheese lover? A group collaborating at BioCurious and Counter Culture Labs is working on making a vegan cheese engineered from the DNA level, no animals needed, except perhaps some DNA from the narwhal. And if you're a computer programmer, stop on by because we want to make DNA the next big programming language. At BioCurious, we're all getting together and being curious about biotech.

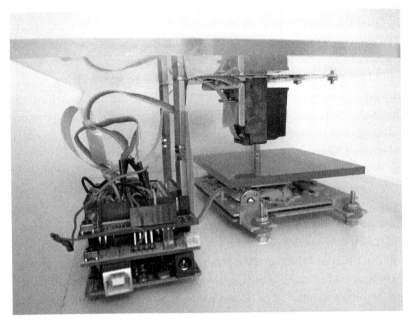

**Figure 12-3.** *The DIY BioPrinter project at BioCurious prints in cells. Photo credit: Patrik D'haeseleer.*

Simply put, BioCurious is a community of people like you who are curious and inquisitive. For everyone who visits, some aspect of biotechnology caught their attention, whether it's cars, food, engineering, programming, or something else. The curious visit to see what all the commotion is about and end up staying.

We first started meeting in Eri Gentry's garage. When the group got too big to fit, six of us went on Kickstarter and raised $35,000 to rent space for a real biotech lab. It's like TechShop, for biology. Since opening in 2011, BioCurious has been lucky enough to be home to many amazing people pursuing their curiosity about biotech, whether they are brand new to biology or work professionally as a scientist.

When we started, our burning questions were all technical. Can we buy lab equipment with our minuscule budget? Can we make scientific discoveries with minimal resources? Is it even legal to open a community biotech lab? Yes, yes, and yes, we quickly learned. But as BioCurious has grown, it's become clear that there are many opportunities beyond the technical.

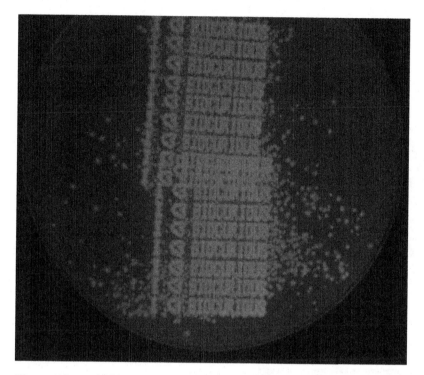

**Figure 12-4.** *A bioluminescent print. Photo credit: Patrik D'haeseleer.*

So, what's next? Community biotech labs are heating up. But it's a mistake to measure the success of a community lab by the number of startup companies created or scientific papers published. I think there's a much greater opportunity than simply re-creating a new university lab/startup incubator model. Today, the biggest opportunity for BioCurious is pushing beyond the technical and into the social elements of science. Can we change who makes scientific discoveries? Can we expand the global conversations around new discoveries to include more groups? Can we increase the number of people who have set foot in a biotech lab?

There are community biotech labs starting all over the world. Soon there will be a community biotech lab in your town! (See DIYbio's website (*http://diybio.org/ local/*) for a full list of community labs worldwide.) Maybe you've got something you're passionate about, or maybe you're still searching. Come check out what's going on in the world of DIYbio, OpenPCR, and BioCurious!

Eri Gentry is cofounder and President of BioCurious, the world's first hackerspace for biotech, and Research Manager at Institute for the Future, a futures think tank. In 2013, Eri was named a White House Champion of Change for Citizen Science and made the Techonomy Top Ten list by nudging out Additive Manufacturing. Photo credit: Eri Gentry.

Tito Jankowski is cocreator of OpenPCR, the DIY DNA copier, and cofounder of BioCurious, the world's first hackerspace for biotech. Read more at his website (*http://titojankowski.com*). Photo credit: Tito Jankowski.

# 13

# Interview: Freaklabs'
# Chris "Akiba" Wang

WRITTEN BY JOHN BAICHTAL

*Akiba talks about moving out of the Tokyo HackerSpace and into the Hackerfarm.*

**John Baichtal: You were at Tokyo HackerSpace, and then...how long were you there? Are you still a member?**

Chris Wang: Nah, not really. I can't make it out there. I live an hour and a half outside of Tokyo now. It was me and another founder of Tokyo HackerSpace, and then a rice farmer. He also was in Tokyo HackerSpace in the beginning, too. The three of us got together and started a farm, actually. My friend was a rice farmer out here. At that time, I was into cheesemaking. I was like, "Well, OK. It'd be nice to have access to raw milk or cheese."

At first, it started kind of innocuously and he's like, "The rent's cheap out in the countryside." So I was like, "Oh, well how cheap?" So right now, I'm renting three buildings and the rent is like $400 a month. Three buildings and then you're basically gifted farmland. It's like, "Here, here. If you can work it, just work it."

**John: And, are you?**

Chris: Not yet. Right now there are still renovations that need to be redone because the place is a bit run down. It's interesting because it went from, "Oh, it'd be kind of a nice thing to have," to, "Wow, OK." If we get a lot of space then, of course, three HackerSpace founders, we're like, "We need all the hackerspace!"

It's interesting because now it's like a hackerspace with a context. It's in a farm context. So, all the projects are going to lean towards agriculture. The other founder, actually, he's a chef. So he's interested from the food angle, kind of organic food, traceable food, just from that side, and food safety. And the rice farmer, he's interested from the agricultural and sustainability aspect. I'm interested in this be-cause I'm interested in environmental monitoring.

It went from some stupid selfish beginnings. Since we're doing a hacker-space, right now the concept that we're going towards is a live-in hackerspace. So people can just come and hang out and crash here.

**John: Do some chores?**

Chris: Yeah, yeah. Contribute where you can. The thing is the cost of living is so cheap that it's like people would pay almost nothing or nothing, right? They'd just do work in kind. Then they can just use the laser cutter. The problem in Tokyo is there are a lot of artists and designers. It's a really tough life if you're just a pure artist or designer without doing a lot of side jobs. It's always the problem of liv-ing in this city. The idea is to have a lot of the advanced tooling out here that they can use, like the designers or actually just people that don't fit into the normal corporate world. They can just crash here and finish their project without having to worry about living cost and stuff.

**John: It's like a residency, almost.**

Chris: It's like a residency. At least at the moment, you don't really need to apply for it.

**John: You just show up?**

Chris: Yeah. If it becomes a problem, then we'll probably figure something out. At the moment, I think it's just a beautiful area.

**John: Is this the next stage of the hackerspace revolution? Instead of having the urban warehouse, have the rural barn that is dirt cheap but still has all the same tools?**

Chris: I forgot to mention, the most important part is here on the farm I have a 200 Mbps fiber-optic line. That's the most important part. I mean, for a hacker-space, you need tools and time. Those are the two main things from what my understanding of hackerspace is. Because you need tools to do interesting projects and you need time. Like, at Tokyo HackerSpace, the problem was always the rent was too high and everyone was working so hard to take care of their own life that they rarely had time to do a lot of projects.

**6699**

The other founder, actually, he's a chef. So he's interested from the food angle, kind of organic food, traceable food, just from that side, and food safety. And the rice farmer, he's interested from the agricultural and sustainability aspect. I'm interested in this because I'm interested in environmental monitoring.

**John: So you're talking about environmental monitoring. You first started doing that with the Fukushima project, right? Or were you into it before?**

Chris: I was into it before. My thing is wireless sensor networks. That was what they used to call the Internet of Things.

**John: With your Chibi Boards, and new stuff that you've developed, you were pretty much ready to go as soon as Fukushima hit. You knew what you had to do?**

Chris: Yeah, yeah. Fukushima was probably the easiest problem to tackle at the time. It was well defined. It was almost like a pure technology issue because you had to send something that is invisible to the human eye. Whereas, the other issue requires huge logistics, like the tsunami victims. Just like logistics and how to get food when the infrastructure is broken down. There are all these other weird problems.

At that time, we had the big meeting. It was on the Tuesday after everything kind of went down. And then we just listed like 10 things that had to be done. We just divide this up and tackle all 10 of them. I think eight of them were complete failures. And one of them turned into Safecast. That one went well. Another one was semi-successful. We were helping with connecting people that had food or supplies with the people that could move them to the areas that needed them.

**John: Right. And you had your jars too, right? The Kimono Lanterns...**

Chris: Oh, yeah! The lanterns. That's actually another interesting story. That lantern, the first version, we sent out the first version for the tsunami victims. It was pretty awesome, and we got a lot of help from a lot of different hackerspaces. And then after that everything kind of died down. And then it kind of got reborn when I got contacted by a lady doing a mission out in Rwanda from NoiseBridge. She contacted me and asked me about solar lanterns, because she thought she could use the Kimono Lantern. But we talked about it and we kind of hashed out a spec on what she really needed. That became the second version of the lantern.

So the original PCB was like a big brown round one, and then it got shrunk down to a super tiny size. But the interesting thing was I started researching it and then I located the actual chip that all the cheap Chinese solar lantern companies use.

**John: I know you can get those for dirt cheap at the end of the season. And there's cool stuff in there, like an RGB LED, like this crystal globe that fits around it and stuff.**

Chris: Yeah, I was always wondering how they can do it so cheap. Because I knew from the lantern that it was tough. And so there's actually a chip, and the chip costs almost, I don't want to say fractions of a penny, but somewhere in that range. And it has all the functions. What it does is, that chip, they use cheap LEDs and they overdrive them. They spike the current. So they just give spikes. It doesn't give a full DC out. They overdrive it. And as it dims back down, they hit it again. It's almost like a PLDM [power LED driver module]. So it looks like it's brighter than it really is. And they also use a lot less power. It's ingenious. So anyway, yeah, I discovered the chip, and I got the BOM cost down to almost nothing. And then the Africa project, the Rwanda project fell through. But now we're going to be using them in the Himalayas for another project. It's kind of weird that everything just kind of moves from project to project.

**John: Right. So tell me about the project you did recently, what was it in, Dharamsala? Where you went to do environmental monitoring?**

Chris: I think it's really interesting. And I think it's going to turn into something kind of cool. So, last year I was asked by some people at UNESCO to go out to Dharamsala, in India. That's basically where the Tibetan community in exile are. It's almost like all the refugees from Tibet went to Dharamsala, India. And they have a huge Tibetan community out there.

I was out there teaching a workshop on wireless sensing. And also wireless sensing using data. At first, I didn't know what to expect, my first time in the Himalayas. I went out and it was really cool. I stayed at this place called Tibetan Children's Village, which is almost like an orphanage for a bunch of Tibetan kids. It was really nice. Last year, I was already kind of planning I want to come back again and maybe do more workshops that are open to the public, since that was kind of an invite-only thing. It was me and I was going to invite my friend from MIT Media Lab. She works with bunny on ChibiTronics, the sticker LEDs.

So we were going to teach kids workshops on technology. But then, CCC [Chaos Computer Club] suddenly got involved too. I was talking to one of the girls that organizes the annual congress. She's actually living out in India, and then she was in Dharamsala, and said, "We should do it together." And I was like, "Yes, CCC!"

And then after that all these other groups started coming in, it was getting kind of weird. It was getting big. And then, other people from MIT Media Lab are involved too. It's like this weird effort. The more interesting side of it is that there's no budget, so there's almost no money. I took on a consulting job. I'm just doing a bunch of consulting jobs and buying junk laptops from the junk bins and then refurbishing them for all the hands-on workshops. It's interesting how this project is spawning a lot of smaller projects.

I think it's going to be cool. The idea is to put together, not just to have one event, but to have an extremely hands-on event. It's going to be two weeks. Right now I'm trying to get two projectors, all identical laptops, two projectors, and two P.A. systems, so events in the future could happen easily.

There are actually three components to this event. The first component is going to be technology and education, and that's what most of the Media Lab people are interested in. They're going to be working with, I guess, the adults, and also the kids on education technology projects. The second track is going to be technology and ethics. We're going to get a bunch of people, especially from CCC out there, especially how ethics needs to be involved, especially with everything that's going on. And the third is on security for activists. The reason why we have the last two tracks is because Dharamsala is like the most hacked place on earth. They're isolated in the Himalayas, but they're under constant attack by a huge variation of grades of hackers, all the way from script kiddies to military.

**John: Just because they're seen as rebelling against the Chinese?**

Chris: I shouldn't say it on the record. But everyone knows who's hacking. The office of the Dalai Lama and Central Tibetan Authority is there. They're seen by China as trying to be separatist, to try and break away from China. The thing is, they don't just hack into the Central Tibetan Authority computers, the whole area, everybody is basically under surveillance.

Once you get on a network, they're going to try and start getting on your computer and identify who the activists are, what the networks are, and basically set up all the monitoring. The issue is, there's such a huge density of activists then security for activists is a really big topic for big event.

**John: So let's get back to the farm, and where you were before you went to the farm. You were living in Tokyo. What made you decide to go to the country? Is it just because it was cheaper? You didn't have to hold down a day job, or something?**

Chris: Well, there were actually a lot of reasons. The Hackerfarm project started about almost two years ago. I just moved out here about five months ago. For almost a year and a half we just had this place, and the idea originally was to try and create a space, like kind of a happy space, on our farm. Especially to promote food projects, like food safety and sustainable agriculture. That was going on for a long time; it just never really moved anywhere. At that time, I was just getting busy with a bunch of projects. Last year I started doing a lot of projects, but most of the projects wouldn't be paid projects. The projects are really interesting, but there was no way to make a living. And I reached one point where it was like, "Yeah, I'm hitting the threshold." I either need to make a decision to get a job or I need to figure out what I can do to cut all my expenses. So I was like, "Oh, fuck it, I'm going to move to the farm." I moved out here, it was crazy. It was the best decision I've ever made.

**John: Nice. Were you freelancing full-time before you moved out?**

Chris: It was kind of like odd jobs. Almost everybody I know, especially in open source hardware, do some kind of strange mix of consulting, workshop, and other random jobs to earn money. I was consulting, pretty light consulting. Most of my money I think I was making off the web shop. But then I hit a patch of, I think, like two or three months the end of the summer season, for some reason, for my shop is a huge lull. But that lull lasted about two to three months and at the same time all the consulting suddenly dried up too. So you just start to see the money go down. Well I was like, "Well, the trajectory is pretty steep." So at that time I had to make some snap decisions. It was like, "OK, it's not going to work."

**John: So are you able to live off of your webstore now that you're at the farm?**

Chris: Yeah. So first of all, after the shop picked up, because it was a temporary lull, and I think towards the end of summer because everyone is going on vacation or coming back from vacation. But then, around November or December when I actually moved out here, my shop traffic actually picked up and I was earning a fairly decent income at that time. And then, also, I got a fat consulting gig. And I was living on the farm where I had almost no living cost. I was like, "Wow. OK. I feel like I'm rich." How did this happen?

**John: You have the freedom to do the fun projects that don't pay, like the Himalayas project.**

Chris: Yeah. Well now, it's interesting because I'm working on the project out in the Himalayas. That's probably my biggest one right now. Like Wrecking Crew doesn't really pay anything, but it's just fun to work with them.

But after I came out here, it was interesting because the cafe suddenly closed down. And the cafe was like the center of the small community out here. Everybody knows each other and there's only one community center, which would be the cafe. And then it closed down, and I was like, "Oh, well that sucks that they closed down. Now there's nowhere to hang out." They'd have like a nice market almost every week, and you just go up there and everybody's selling their food. Everybody grows their own food, so they just kind of buy it out of being nice to their friends. It was really cute. So when that shut down, I was talking to a friend, and I was like, "Why don't we extend the Hackerfarm to the cafe?" So now we're doing talks in the cafe.

But then the cafe has these huge grounds on the outside. It's on like a huge, almost like an acre lot, of property. Since we have a lot of outdoor space, we're going to set up a theater, like a live stage, and we're going to invite performers out. So the next thing is that everybody goes to bed early out here. So you can't really have outdoor performances at night. So we're going to experiment with silent theater. So you transmit wirelessly, everybody wears headphones.

**John: Oh OK, so you don't disturb the farmers who go to sleep at night.**

Chris: Yeah, yeah. So it's like, now it's starting to sound interesting.

**John: All right.**

Chris: It's pretty fun. I have to say, when you get rid of all of the financial anxiety and stuff and you can just focus on interesting projects, and there's just so many. It's just unbelievable.

**John: So, this farm gives you the freedom to do the things that you want to do without having to be slaving away at jobs all the time. You're able to use your webstore and your freelancing to make some money, but your cost of living is so low that you can do stuff like the Dharamsala project and your experiments with farming and starting a cafe. That's something that you couldn't have done if you were working 50 hours a week in a city.**

Chris: Yeah, yeah. I think the only other thing I would add is it's so awesome and that's why we're trying to bring other people out here. One thing that really kind of pisses me off is designers who are really good but they can't earn a living doing what they do. It's the same problem with dancers and performing artists too. It'd be kind of cool to try and right now I guess the idea is to try and attract creative people out here, where they can just be creative without having to worry about living costs. When they want to display it, or whatever, they just take a bus into Tokyo. So that's the idea. But I guess the archetype would be just kind of unfettered creative freedom.

Chris "Akiba" Wang is fascinated by all things electronic, but mostly about the potential for wireless and wireless sensor networks to make huge changes in how everyday things are done. He's also very interested in open source software, open source hardware, and the recent phenomenon of micro-manufacturing. Photo credit: Chris Wang.

John Baichtal has written several Maker-themed books, including *Arduino for Beginners* (Que), *Make: Lego and Arduino Projects* (Maker Media), *Hack This: 24 Incredible Hackerspace Projects from the DIY Movement* (Que), *Maker Pro* (Maker Media), and *Robot Builder: The Beginner's Guide to Building Robots* (Que).

Photo of John Baichtal (left) racing his Pinewood Derby car. Photo credit: Marie Flanagan.

# 14

# Making Things Is Even More Vital Than You Think

WRITTEN BY **DAVID GAUNTLETT**

Hello, Makers! I might be an odd-one-out in this book, because I am not a professional Maker. I do write and research about Makers and making as part of my work, but otherwise I am a thoroughly unprofessional Maker. I don't make things super-properly. But I do make things. I make things because you learn stuff, you link with others, and because it's fun.

If you're a professional Maker, or you want to be one, the amateur Makers like me are not your opposite, and we are not your competition. We are, potentially, your best fan base, because we know the pleasures and pains of making. If you're doing really good stuff, we will know it, we will recognize it, and we will share it.

The Maker Movement, which is essentially—or, I think, most excitingly— about the rise of networks of making enthusiasts, rather than professionals, is still great news for professional Makers. These are the people that know the value of the handmade, who recognize the passion and creativity that goes into designing and making something that wasn't there before. And these are the people who really believe that the best things in life don't come ready-made from factories.

I think making is especially powerful because it is about connecting. I even wrote a book called *Making is Connecting* (Polity) so that I could explore this idea in full. The starting point was that making is connecting in three significant ways:

- Making is connecting because you have to connect things together—materials, ideas, or both—to make something new.
- Making is connecting because acts of creativity usually involve, at some point, a social dimension, and connect us with other people.
- Making is connecting because through making things and sharing them, we boost our engagement with the world, and build connections with cultures—social, political, or artistic—and the environment.

So making things is certainly nice and rewarding and fun, but it isn't only nice and rewarding and fun. It's also quite crucial to our future on this planet, because it builds people's sense of connectedness and belonging and having a stake in things. And we need to nurture and inspire generations of hands-on learners and creative people, partly because they bring diverse ideas and novel innovations to our culture, and partly because without creative people we haven't got anybody to solve the ecological and social challenges that lie ahead.

And when you make things yourself, you break that expectation, you step into the world more actively. So I think the most important thing is taking that step. It doesn't matter what you've made, or whether it's as good or effective or neat as something made by someone else, or a company. The important thing is that you've made something and put it into the world.

Here I'll make six points which follow from those observations.

## 1. Why make things?

As part of the research for *Making is Connecting*, I pulled together a number of existing studies of people's motivations for making things. I looked at research about those doing it online—bloggers, YouTube video-makers, and other sharers of digital creativity. I also looked at studies of hands-on people doing it offline—

such as crafters, robot-builders, and makers of jewelry and clothing. It turned out that all of these kinds of people had much in common.

It boiled down to three things:

*Agency*

On an individual level, making things offers opportunities for pleasure, thought, and reflection, and helps to cultivate a sense of the self as an active, creative agent in the world.

*Community*

Making is also, crucially, a social activity—people spend time creating things because they want to be an active participant in dialogues and communities.

*Recognition*

There was also a desire to be noted and respected for what one has made, and for the contribution made within a community of interesting (like-minded) people.

The first two might not seem especially surprising, but the third one sticks out —people want recognition? But of course they do—and there's nothing wrong with that. Human beings need to be creative in everyday life—bringing variety and invention to one's own world is one of the things that makes life worth living—but we also like to know that we've made a positive difference to others.

So being part of a community that notices and respects your particular kind of expertise is really important for people. A pleasing aspect of how the Internet connects up Makers is that way in which you might have expertise that does not happen to engage the people around you, but which can always find an appreciative audience somewhere in the world. And on a practical level, of course, the same thing helps professional Makers find people who love, and want to buy, their work.

## 2. Spirit of the Maker

One source of inspiration for me is the Arts and Crafts movement, which took off in the late nineteenth century, but was built upon a set of ideas that are still incredibly relevant. Emerging as a response to mass-produced goods, the Movement was all about the power of making distinctive, expressive things. A lovely bit of Arts and Crafts philosophy comes from John Ruskin, who expressed strong admiration for roughly-made, characterful things, such as the gargoyles on medieval cathedrals. Ruskin alienated himself from the Victorian art establishment with his passion for such quirky, unfinished, unprofessional things—but Ruskin's point was that these formal qualities are not what we should really value: the important thing is that you can see in the created object the spirit of a Maker who feels an urge to communicate, to express themselves, to say something or have an impact on others.

If we carry that idea forward to today, we can see that the rise of DIY social media—like homemade YouTube videos—and contemporary craft, and the Maker Movement, are all to do with people wanting to make something, something that is uniquely their own. Each thing celebrates the spirit of the Maker, it reflects the individual who made it, and celebrates the power of all of us to make creative choices, and to build the world around us. Every made thing serves as least two roles, being nice in itself, but also by inspiring other people to make other things. It shows that we don't have to only consume mass-produced goods made by professional industries, but that we can make culture ourselves, anew, each and every day. Which brings us to the third point.

## 3. Make your own luck

The surge in all kinds of everyday creative activities—that we see in the Maker Movement, crafts, blogs, YouTube—is connected to a DIY ethos where people are realizing the power of "just doing stuff" (as the title of a book by Rob Hopkins suggests; I am also grateful to the designer Kate Moross for the title, "Make your own luck").

Crowdfunding websites such as Kickstarter, and online sales platforms such as Etsy—especially when used in conjunction with other social media, like Twitter —have helped people to gain attention and support for their creative work. In 2008, Kevin Kelly developed the heartening argument about "1,000 True Fans," which suggests that a creator "needs to acquire only 1,000 True Fans to make a living." The idea is that if you can get 1,000 followers who are willing to spend $100 each year on what you do, then you'll have enough to make a decent living ($100,000, minus costs). This is appealing: 1,000 fans doesn't seem too many— it's a number you can picture. And social media tools make it easier to build a conversational kind of relationship with fans up to a certain number—say, 1,000.

Of course, alas, if you think about it, it all unravels rather quickly: the fans are paying money for things that cost money to produce, create, or perform, so the profit kept by the artist is usually a small fraction of this amount—out of the $100,000 spent by fans on your stuff, you might be lucky to keep $10,000, and that's not a great wage. Also, having 1,000 super-committed fans of that sort probably means you need a wider pool of tens of thousands of really-rather-interested fans. And maintaining that level of interest over time is incredibly hard work as well. Kelly acknowledged these objections in later blog posts. There's still a DIY appeal to it, though, and the argument is more persuasive with digital products—books, music, or apps—which can be created, copied, and distributed at quite low cost, while still requiring time and talent to make them. Handmade physical items are at the opposite end of the spectrum, of course, requiring time and talent and materials to make just one of them.

Kelly's vision was that rather than having a small number of megastars, we might be able to shift to a more diverse creative culture, where a wider range of inventive artists would be able to build a sustainable life. Although the model, on inspection, looks difficult and precarious, a lot of comments on Kelly's blog posts referred to the pleasure of control over an artistic career, and "making a living" from it, with a meaningful connection to some people who love the work, even if the artist is not having big hits or mainstream success.

With social media enabling creative people to raise awareness of what they do, build communities around shared interests, and enthuse and inspire each other, the broader point is that you can make your own luck without waiting for "media attention" or paying for expensive advertising. It's still, like many things, quite hard work, but it's something you can do right now, and there are many examples of people making a success of it.

## 4. Small steps into a changed world

Making might be a joy but is also crucial to our future, and a vital social movement, then it can start to seem like an intimidatingly big deal. But it's something we can all do, in small steps. And those small steps are really important.

If you've decided to be a professional Maker, you're taking a big step, but hopefully it's not a step away from everyday, amateur making—on the contrary, your big step can inspire important smaller steps in others. When someone knits a hat, writes a song, or shares a video, it's easy to see these things as trivial in terms of changing the world. But the small steps are incredibly important because they all add up.

When you make something and put it into the world—as I've said in the third "making is connecting" point—I think it changes your relationship to the world, to your environment, the people around you, and your relationship to the stuff in the world. In everyday life, we're expected to be "participating" in the world, but in a particular kind of way—basically using stuff made by other people, and consuming or being fans of things made by other people.

And when you make things yourself, you break that expectation, you step into the world more actively. So I think the most important thing is taking that step. It doesn't matter what you've made, or whether it's as good or effective or neat as something made by someone else, or a company. The important thing is that you've made something and put it into the world. So you're making your mark, and you've taken that active step. You're making a difference. It's fine if it's a tiny difference, or if it's only noticed by one person. It's the step you've made, and so, it's a great step.

## 5. Passion

The word "passion" is undoubtedly overused these days—you can find people, as part of their employment, claiming to be passionate about mortgages, motor insurance, and all kinds of things. But the rapturous engagement in ideas, materials, and communities of creativity enjoyed by enthusiastic Makers can reasonably be called a kind of passion, and it's this passion that really drives the whole rise of Maker culture. The fascination with what you can do next, and how other people will engage with it, is the reason why we try and try again, and is the motivation which supports the difficult and precarious mission to build that following of "1,000 true fans"—or more!—rather than signing up to a more normal job.

The striking thing about the Maker Movement is that it's always people doing things because they want to. It's the passionate engagement of Makers that brings it all to life, and drives everything. They're not usually doing it for qualifications, status, or money—making stuff is not normally the quickest route to these things—or because someone told them to, or it's part of their job. It's all just because it's what they want to do.

## 6. One thing leads to another

All of this has good knock-on domino effects. Many of us lament the lack of emphasis on playful creativity and experimentation in our education system. We want more creative schooling for kids, with more hands-on learning, making stuff, trying things out, and less emphasis on tests of remembered knowledge. And yet we feel powerless to change it—it all seems so big and ingrained.

I've been at conferences and committees about the need to change education in this way, and the focus is often on children and schools. That makes sense, of course, on the face of it—we are, after all, talking about the education of children, in schools. But actually these are just elements within a system. If you want to have a culture of playful learning and experimentation, you need adults to embrace a culture of playful learning and experimentation before you can expect that we might try to make it happen in schools.

I've had the curious experience, recently, when making this point, that people seem to take this argument on board, but then it turns out that they thought I meant parents. But I'm not just saying that we want parents to agree with the notion of a more playful education for their kids—it's much bigger than that. I'm saying we need adult culture itself to become more playful and creative, because only then will that really be seen as something valuable that we can hand on to children. We have to look at culture as a whole system, and not think that the "education" bit can be separated off and fixed without changing the rest of it.

Now, this means I have upped the ante dramatically—how are we meant to change a whole culture? But cultures change little by little, on the terrain of everyday life. And so what Makers are doing is modeling the ways in which life can be transformed for others.

I use "modeling" here in the sense that I learned from the psychologist Mark Runco, where a kind of everyday engagement in creative practice offers an inspiring model to other people—not that you would copy the creative behavior itself, but rather that you might be inspired to do more creative things yourself, in everyday life, because you have seen someone else happily engaged in doing creative things in everyday life. In this way, Makers are changing the future, just by doing what they like to do, and sharing it with others.

The points I've listed here are just a few of the many possible reasons for celebrating Makers and making. People have made stuff for many thousands of years. We build meanings, connections, and communities around the things we make. Having started off by saying that making is socially vital and not just good fun, I should end by twisting that around. Making stuff might be part of a grand social movement, but on the individual level, most importantly of all, making stuff builds relationships, binds people together, and can be a source of great happiness.

David Gauntlett is a professor in the Faculty of Media, Arts and Design, University of Westminster, UK. His teaching and research is about self-initiated everyday creativity, and cultures of making and sharing. He is the author of several books, including *Creative Explorations* (Routledge), *Making Is Connecting* (Polity), and Making Media Studies (Peter Lang, will be published in 2015). He has worked with a number of the world's leading creative organizations, including the BBC, the British Library, and Tate. For almost a decade he has worked with the LEGO Group on innovation in creativity, play, and learning. Photo credit: David Gauntlett.

# 15

# Soylent Supply Chain

---

Written by **Andrew "bunnie" Huang**

---

The convenience of modern retail and ecommerce belies the complexity of supply chains. With a few swipes on a tablet, consumers can purchase almost any household item and have it delivered the next day, without facing another human. Slick marketing videos of robots picking and packing components and CNCs milling components with robotic precision create the impression that everything behind the retail front is also just as easy as a few search queries, or a few well-worded emails. This notion is reinforced for engineers who primarily work in the domain of code; system engineers can download and build their universe from source —the FreeBSD system even implements a command known as `make build world`, which does exactly that.

The fiction of a highly automated world moving and manipulating atoms into products is pervasive. When introducing hardware startups to supply chains in practice, almost all of them remark on how much manual labor goes into supply chains. Only the very highest volume products and select portions of the supply chain are well-automated, a reality which causes many to ask me, "Can't we do something to relieve all these laborers from such menial duty?" As menial as these duties may seem, in reality, the simplest tasks for humans are incredibly challenging for a robot. Any child can dig into a mixed box of toys and pick out a red $2 \times 1$ Lego brick, but to date, no robot exists that can perform this task as quickly or as flexibly as a human. For example, the KIVA Systems mobile-robotic fulfillment system for warehouse automation (*http://www.kivasystems.com/resources/demo*) still requires humans to pick items out of self-moving shelves, and FANUC pick/pack/

pal robots can deal with arbitrarily oriented goods, but only when they are homogeneous and laid out flat. The challenge of reaching into a box of random parts and producing the correct one, while being programmed via a simple voice command, is a topic of cutting-edge research.

**Figure 15-1.** *bunnie working with a factory team. Photo credit: Andrew Huang.*

The inverse of the situation is also true. A new hardware product that can be readily produced through fully automated mechanisms is, by definition, less novel than something which relies on processes not already in the canon of fully automated production processes. A laser-printed sheet will always seem more pedestrian than a piece of offset-printed, debossed, and metal-film transferred card stock. The mechanical engineering details of hardware are particularly refractory when it comes to automation; even tasks as simple as specifying colors still rely on the use of printed Pantone registries, not to mention specifying subtleties such as textures, surface finishes, and the hand-feel of buttons and knobs. Of course, any product's production can be highly automated, but it requires a huge investment and thus must ship in volumes of millions per month to amortize the R&D cost of creating the automated assembly line.

Thus, supply chains are often made less of machines, and more of people. Because humans are an essential part of a supply chain, hardware makers looking to do something new and interesting oftentimes find that the biggest roadblock to their success isn't money, machines, or material: it's finding the right partners and people to implement their vision. Despite the advent of the Internet and robots, the supply chain experience is much farther away from Amazon.com or Target than most people would assume; it's much closer to an open-air bazaar with thousands of vendors and no fixed prices, and in such situations getting the best price or quality for an item means building strong personal relationships with a network of vendors. When I first started out in hardware, I was ill-equipped to operate in the open-market paradigm. I grew up in a sheltered part of Midwest America, and I had always shopped at stores that had labeled prices. I was unfamiliar with bargaining. So, going to the electronics markets in Shenzhen was not only a learning experience for me technically, it also taught me a lot about negotiation and dealing with culturally different vendors. While it's true that a lot of the goods in the market are rubbish, it's much better to fail and learn on negotiations over a bag of LEDs for a hobby project, rather than to fail and learn on negotiations on contracts for manufacturing a core product.

**Figure 15-2.** *One of bunnie's projects is Novena, an open source laptop. Photo credit: Andrew Huang.*

This point is often lost upon hardware startups. Very often I'm asked if it's really necessary to go to Asia—why not just operate out of the US? Aren't emails and conference calls good enough, or worst case, "can we hire an agent" who manages everything for us? I guess this is possible, but would you hire an agent to shop for dinner or buy clothes for you? The acquisition of material goods from markets is more than a matter of picking items from the shelf and putting them in a basket, even in developed countries with orderly markets and consumer protection laws. Judgment is required at all stages—when buying milk, perhaps you would sort through the bottles to pick the one with greatest shelf life, whereas an agent would simply grab the first bottle in sight. When buying clothes, you'll check for fit, loose strings, and also observe other styles, trends, and discounted merchandise available on the shelf to optimize the value of your purchase. An agent operating on specific instructions will at best get you exactly what you want, but you'll miss out better deals simply because you don't know about them. At the end of the day, the freshness of milk or the fashion and fit of your clothes are minor details, but when producing at scale even the smallest detail is multiplied thousands, if not millions of times over.

Because humans are an essential part of a supply chain, hardware makers looking to do something new and interesting oftentimes find that the biggest roadblock to their success isn't money, machines, or material: it's finding the right partners and people to implement their vision.

More significant than the loss of operational intelligence, is the loss of a personal relationship with your supply chain when you surrender management to an agent or manage via emails and conference calls alone. To some extent, working with a factory is like being a houseguest. If you clean up after yourself, offer to help with the dishes, and fix things that are broken, you'll always be welcome and receive better service the next time you stay. If you can get beyond the superficial rituals of politeness and create a deep and mutually beneficial relationship with your factory, the value to your business goes beyond money—intangibles such as punctuality, quality, and service are priceless.

I like to tell hardware startups that if the only value you can bring to a factory is money, you're basically worthless to them—and even if you're flush with cash from a round of financing, the factory knows as well as you do that your cash pool is finite. I've had folks in startups complain to me that in their previous experience at say, Apple, they would get a certain level of service, so how come we can't get the same? The difference is that Apple has a hundred billion dollars in cash, and can pay for five-star service; their bank balance and solid sales revenue is all the top-tier contract manufacturers need to see in order to engage.

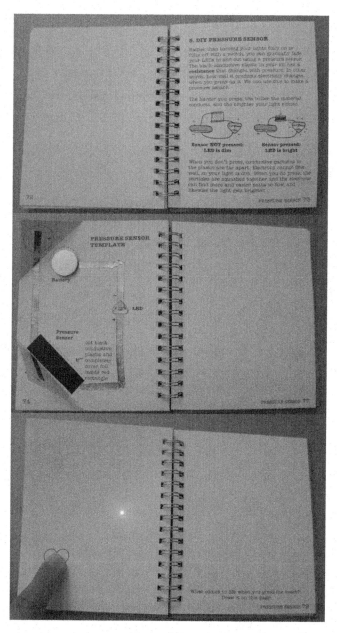

**Figure 15-3.** *Circuit Stickers, adhesive-backed electronic components, is another of bunnie's projects. Photo credit: Andrew "bunnie" Huang.*

On the other hand, hardware startups have to hitchhike and couch-surf their way to success. As a result, it's strongly recommended to find ways other than money to bring value to your partners, even if it's as simple as a pleasant demeanor and an earnest smile. The same is true in any service industry, such as dining. If you can afford to eat at a three-star Michelin restaurant, you'll always have fairy godmother service, but you'll also have a $1,000 tab at the end of the meal. The local greasy spoon may only set you back ten bucks, but in order to get good service it helps to treat the wait staff respectfully, perhaps come at off-peak hours, and leave a good tip. Over time, the wait staff will come to recognize you and give you priority service.

At the end of the day, a supply chain is made out of people, and people aren't always rational and sometimes make mistakes. However, people can also be inspired and taught, and will work tirelessly to achieve the goals and dreams they earnestly believe in: happiness is more than money, and happiness is something that everyone wants. For management, it's important to sell your product to the factory, to get them to believe in your vision. For engineers, it's important to value their effort and respect their skills; I've solved more difficult problems through camaraderie over beers than through PowerPoint in conference rooms. For rank-and-file workers, we try our best to design the product to minimize tedious steps, and we spend a substantial amount of effort making the tools we provide them for production and testing to be fun and engaging. Where we can't do this, we add visual and audio cues that allow the worker to safely zone out while long and boring processes run. The secret to running an efficient hardware supply chain on a budget isn't just knowing the cost of everything and issuing punctual and precise commands, but also understanding the people behind it and effectively reading their personalities, rewarding them with the incentives they actually desire, and guiding them to improve when they make mistakes. Your supply chain isn't just a vendor; they are an extension of your own company.

Overall, I've found that 99% of the people I encounter in my supply chain are fundamentally good at heart, and have an earnest desire to do the right thing; most problems are not a result of malice, but rather incompetence, miscommunication, or cultural misalignment. Very significantly, people often live up to the expectations you place on them. If you expect them to be bad actors, even if they don't start out that way, they have no incentive to be good if they are already paying the price of being bad—might as well commit the crime if you know you've been automatically judged as guilty with no recourse for innocence. Likewise, if you expect people to be good, oftentimes they will rise up and perform better simply because they don't want to disappoint you, or more importantly, themselves. There is the 1% who are truly bad actors, and by nature they try to position themselves at the most inconvenient road blocks to your progress, but it's important to remember that not everyone is out to get you. If you can gather a syndicate of friends large enough, even the bad actors can only do so much to harm you, because bad actors still rely upon the help of others to achieve their ends. When things go wrong your first instinct should not be "they're screwing me, how do I screw them more," but should be "how can we work together to improve the situation?"

In the end, building hardware is a fundamentally social exercise. Generally, most interesting and unique processes aren't automated, and as such, you have to work with other people to develop bespoke processes and products. Furthermore, physical things are inevitably owned or operated upon by other people, and understanding how to motivate and compel them will make a difference in not only your bottom line, but also in your schedule, quality, and service level. Until we can all have Tony Stark's JARVIS robot to intelligently and automatically handle hardware fabrication, any person contemplating manufacturing hardware at scale needs to understand not only circuits and mechanics, but also how to inspire and effectively command a network of suppliers and laborers.

After all, "it's people—supply chains are made out of people!"

---

Andrew "bunnie" Huang is an open source hardware designer, Xbox hacker, and creator of the Chumby. He lives in Singapore. In 2012, he received an EFF Pioneer Award for his work in hardware hacking, as well as his advocacy of the open source philosophy. Photo credit: Andrew Huang.

---

# 16

# Quit Your Day Job

WRITTEN BY **SOPHI KRAVITZ**

Make stuff you love, with people you like...all you have to do is give two weeks notice.

I used to come home from my full-time engineering day job on Friday evenings and flop down on the couch, worn out and depressed. An entire week would have gone by and it wasn't until the end of the week that I'd have space in my head to realize that I hadn't had the time or head space to get into my home workshop to design and build something of my own.

What I really wanted to do was make my own projects. And I'm greedy, not wanting just evenings and weekends, I wanted freedom to work on projects all the time. This is the story of how I got that lifestyle.

A little history: I worked as a hardware design engineer for seven years, mostly designing control systems for industrial equipment. As part of an engineering team, some of the things I worked on were lab glassware washers, freeze-dryers, temperature systems, and gas delivery systems. Basically, these machines are large-scale robots with integrated hardware, software, and electronics. Fun!

I enjoyably switched projects (and jobs) quite often, just to get a reasonable amount of time off. As a designer on a small team with controlling bosses and coworkers, it was difficult to get out of the office for more than a week at a time. Plus, deadlines.

Having worked in a variety of different kinds of jobs and industries, it is safe to say that here in the US, taking more than a week vacation without backlash is hardly the norm.

I visited Guinea, China, New Zealand, and a bunch of Western European countries, never leaving The Company for longer than nine days. Besides traveling to other countries, I was constantly angling to get "Makecations" to work on personal projects (Figures 16-1 and 16-2).

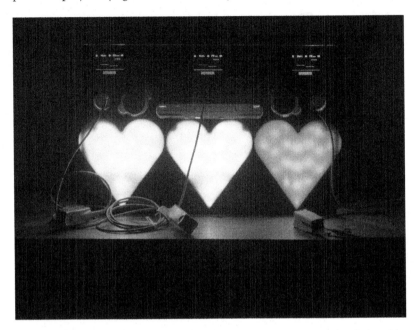

**Figure 16-1.** *The HeartBeat BoomBox utilizes three oximeter sensors designed into a hacked Sony boombox. Three participants control the piece at the same time, causing drumbeats to play when their hearts beat. Photo credit: Sophi Kravitz.*

Naturally, once the team finished our work project we'd be waiting around for something to do. While waiting for management to provide the next project, we'd be pretending to be busy. My requests to go home and wait, unpaid, were always met with disapproval. So I'd move on, taking six weeks off in between.

The upside of having a full-time job, beyond the steady paycheck and health benefits, is that a full-time job can get you collaborators, business partners, and lifelong friends. If you're in a full-time job right now and considering leaving, take advantage of this. Once you're on your own, it can be quite difficult to find technical collaborators, as they tend to be popular with everyone.

By 2010, I was so unhappy having my time and creativity controlled by an employer, I needed to figure out an escape. I would write to friends that I wanted to support myself doing something that was time and location independent. And have enough time and money left over to make projects. The way I saw it, the choices were clear:

1. Be a consultant

2. Get a part-time, permanent job

3. Design a product/sell a product

All of these choices would need some startup time and savings to subsist on while waiting for the new income to catch on.

I have a blog where I interview people about quitting their jobs and specifically what it takes to quit. There is the obvious, knowing what you're going to do when you suddenly gain 40+ hours back per week. And then, most people say they have at least a year of wages saved up before they quit their full-time jobs. An entire year of wages!

I definitely didn't have a huge stockpile of savings and I wasn't making enough money (or likely to in the future) to get a huge stockpile any time soon. Quitting without some kind of income was not going to work for me, and my guess is that it wouldn't work for you either. As it turned out, I have now tried all three of these options and currently mix them all up depending on what projects I want to work on!

An analytical look at the situation made it clear that an immediate solution would be to work part time. At 20% less pay per day off per week, it seemed like employers would be beating down my door. Sadly, this was not the case. As I quickly found out, established companies don't want part-time engineers, they want someone who is going to dedicate all of the five work days a week of your thinking power to them. They want to own you and your creativity. I did manage to get a startup to hire me for a four-day workweek, but this ultimately led to fighting with my boss about schedule, caving in to full-time hours, and being laid off.

Even though I got laid off, working four days a week for someone else seemed almost perfect. Each week, by Thursday afternoons, little political fights were happening, managers were seeking control, and I'd be like, "See ya, have a nice weekend." I thought the only thing better would be to work only three days a week for someone else. Three days of pay wouldn't be much, but since I live in a small town, it would probably be OK.

I switched fields, moving into technical sales and applications because it was customer-driven, rather than boss-driven. With this kind of work, I could work from anywhere via my phone. I could solder up a new project while chatting with a customer and get paid to do it. I would pack all of my meetings into Tuesdays and

Wednesdays and make sales calls first thing in the morning on Mondays. A popular trick to get in touch with important people is to call them first thing on Mondays —it's the anti-slacker hour.

I now had four days completely off per week. Four days in a row without many distractions is long enough to make a lot of projects. This was the first time in my life that I'd had so much time to pursue my personal interests, and it was unbelievable. When I used to only work on projects during the weekends, the first day was spent acclimating and the second day making mistakes. The following weekend might follow the same trajectory, but sometimes I'd skip a weekend due to family stuff and then be so far behind it was hard to regain interest. Having that third or fourth day allowed me to actually get stuff done in a normal amount of time.

From talking to many people about work, my takeaway is that this is what most of us want: time to pursue our interests and enough money to not stress about the time it takes to pursue our own interests—especially if we're not making much money from these interests. The reason we do stuff that doesn't inspire us is simple. Money. We need to support ourselves, our families, and our lifestyles. I meet a lot of people who are miserable in their jobs and not spending their lives doing anything that they love. Many of them tell me that they're stuck, they're stuck due to the economy, they're stuck because they need the money they make, they're stuck because they're supporting a family, they're stuck because they don't know how to get out of the job they have. I think that's awful—all of us are such creative people with so much to offer the world and ourselves, getting the lifestyle we want can and should be a priority.

The path to self-sufficiency is supporting yourself and your family with work that is both time and location independent, and as a bonus, is something that allows you to grow your skills and creativity. We all want to spend our time on stuff that's meaningful to us. If you're a Maker, the projects you want to pursue take a lot of time and if you're not constantly learning, you're probably bored!

After having a part-time schedule for about a year, I'd made some large scale art-engineering projects. This was a huge accomplishment to me as I had done exactly no projects while having a full-time job. It was now time to explore product development.

What we see on Kickstarter and read about companies getting funding is all true. You can start out with a prototype and no money and obtain massive funding. The funding may be enough to support you and your team while you develop your product. That situation isn't true for everyone—your idea may not be that good, your idea may not be that marketable, your idea may take you a year or more to develop, you may not have the skills to develop it and not have anyone to work with, you may not know where to begin to get funding, and so on. It is easy to read the Internet and perceive that everyone lives in startup saturation, but most of us don't. Most people live in regular towns that have only a small percentage of people interested in product design, making stuff, and starting companies.

In my situation, I had an idea for a product, and even a customer. The product was a sensor device for neuroscientists who were using mice in research. The product is called a nosepoke (Figure 16-2), it's a device that provides a sensor for the mice to "poke" their heads into, while sending the "yes, the mouse poked" signal to a data collection system made by a company called Noldus. When I was asked to develop this, a nosepoke product that interfaced easily with Noldus wasn't available. My customer was a scientist who wanted something the lab could work with easily. As is quite normal, the customer didn't pay me to live while I developed it. While working on this part time, it took just over a year to develop, make into a tested and reliable product, and deliver the first batch of 50. And then it took another eight months to get paid. It was a good thing I still had my part-time job while doing this!

**Figure 16-2.** *The nosepoke is used in mouse mazes: a mouse pokes its head into the hole, triggering an IR sensor. Photo credit: Sophi Kravitz.*

A couple of years ago I left my part-time permanent job to create a technical consultancy. This is basically a catchall for doing technical odd jobs. The goal of my consultancy is to pursue interesting paid projects that are location independent and to have enough time left over to pursue unpaid personal projects. I want to have the freedom to try art-engineering projects and research ideas that may never make me a lot of money.

Under the consultancy, I am lucky to do a lot of different things, from electrical design, finding elusive issues in other people's designs (ghost in the machine kind of stuff), blogging, applications/sales, and tutoring. I'm also working on a new product. This lifestyle, while being location and boss independent, is somewhat like being in school, without the structure. Time management and discipline is important, or there won't be any time left over to make stuff!

This lifestyle is definitely hard. I am sometimes out of work for a few weeks at a time. I get all of my work by word of mouth, and sometimes it feels like I've worked for every small business in my small town. Having limited income makes me feel panicked—I have dark thoughts of "will I ever work again?" or worse, "will I be forced into a soul-sucking job designing something that I hate?"

When you are ready to stop working so much in favor of making projects, products, or saving your sanity, out of the three options that are available: part-time, permanent employment, making and selling a product, or having a consultancy, I like working part-time the best. It is clearly the least risky way to quit working full-time. If you can get a part-time job which brings in enough money to support yourself, you'll also enjoy having enough time to pursue making the projects you've always wanted to. You can also use a part-time job to bootstrap your startup, fund your product design, and support you while you figure out what you want to do next!

Sophi Kravitz has been creating interactive works since the completion of her first animatronic project in 2004. Sophi is a formally trained engineer whose first career was in Special FX makeup for film and theater. It was then, while creating works that were seen on small screens or large, that she realized the great fun in creating works that can easily satisfy an audience or group of participants. Although her art-engineering pieces have a technical aspect, such as electronics or code, participants interact with them as simple and beautiful playthings. Photo credit: Sophi Kravitz.

# Index

## Symbols

## A

## B

## C

*We'd like to hear your suggestions for improving our indexes. Send email to index@oreilly.com.*

## About the Editor

**John Baichtal** is a contributor to *Make:* magazine and *Wired's* GeekDad blog. He is the coauthor of *The Cult of LEGO* (No Starch Press) and author of *Hack This: 24 Incredible Hackerspace Projects from the DIY Movement* (Que Publishing).

## Colophon

The cover image is by Riley Wilkinson. The cover fonts are URW Typewriter and Guardian Sans. The text font is Crimson; the heading font is OpenSans; and the code font is Ubuntu Mono.